189

Topics in Current Chemistry

Springer

Berlin
Heidelberg
New York
Barcelona
Budapest
Hong Kong
London
Milan
Paris
Santa Clara
Singapore
Tokyo

Stereoselective Heterocyclic Synthesis I

Volume Editor: P. Metz

With contributions by
G. Kettschau, A. Padwa,
L. F. Tietze

Springer

This series presents critical reviews of the present position and future trends in modern chemical research. It is addressed to all research and industrial chemists who wish to keep abreast of advances in the topics covered.

As a rule, contributions are specially commissioned. The editors and publishers will, however, always be pleased to receive suggestions and supplementary information. Papers are accepted for "Topics in Current Chemistry" in English.

In references Topics in Current Chemistry is abbreviated Top. Curr. Chem. and is cited as a journal.

Springer WWW home page: http://www.springer.de
Visit the TCC home page at http://www.springer.de/

ISSN 0340-1022
ISBN 3-540-62678-6
Springer-Verlag Berlin Heidelberg New York

Library of Congress Catalog Card Number 74-644622

Cover design: Friedhelm Steinen-Broo, Barcelona
Typesetting: Fotosatz-Service Köhler OHG, 97084 Würzburg
SPIN: 10552203 66/3020 – 5 4 3 2 1 0 – Printed on acid-free paper

Volume Editor

Prof. Dr. Peter Metz
Institut für Organische Chemie
Technische Universität Dresden
Mommsenstr. 13
D-01062 Dresden, Germany

Editorial Board

Preface

Heterocycles play a central role in organic synthesis. Above all due to the interesting biological activities associated with a large number of these structurally diverse compounds, many heterocycles have been and will be challenging targets for total synthesis. Moreover, even if the final goal of a synthesis is not heterocyclic, at least a central intermediate or a key reagent used along the synthetic sequence most surely will be. This holds especially true if stereoselectivity is an important issue, as modern heterocyclic chemistry provides the synthetic organic chemist with an excellent arsenal of methods and strategies for the stereocontrolled construction and elaboration (including the cleavage) of heterocycles. Recent years have witnessed exciting new findings in this field, and it is the aim of this two-volume set on "Stereoselective Heterocyclic Synthesis" within the series Topics in Current Chemistry to present a selection of these novel developments.

As the guest editor I am very glad that leading researchers in this area have contributed highly inspiring accounts with up-to-date coverage to this compilation. Part I features chapters on *"Hetero Diels-Alder Reactions in Organic Chemistry"* by *L.F. Tietze* and *G. Kettschau* describing the state of the art for these useful [4+2] cycloadditions, which yield a wide variety of heterocycles and *"Tandem Processes of Metallo Carbenoids for the Synthesis of Azapolycycles"* by *A. Padwa* surveying attractive routes to complex ring systems based upon 1,3-dipolar cycloadditions. Part II comprises chapters on *"Using Ring-Opening Reactions of Oxabicyclic Compounds as a Strategy in Organic Synthesis"* by *P. Chiu* and *M. Lautens* focussing on the preparation and the synthetic utility of the versatile title compounds, *"The Nucleophilic Addition/Ring Closure (NARC) Sequence for the Stereocontrolled Synthesis of Heterocycles"* a powerful tactical combination discussed by *P. Perlmutter*, *"Chiral Acetylenic Sulfoxides and Related Compounds in Organic Synthesis"* by *A.W.M. Lee* and *W.H. Chan* emphasizing the use of sulfur-activated acetylenic and vinyl units for the efficient preparation of heterocycles, and *"N-Sulfonyl Imines – Useful Synthons in Stereoselective Organic Synthesis"* by *S.M. Weinreb* giving a comprehensive review on the chemistry of these valuable electron-deficient compounds.

I hope that the articles collected in this two-volume set on "Stereoselective Heterocyclic Synthesis" will not only serve experts in the field but will also attract the interest of scientists not yet familiar with this fascinating research topic.

Dresden, March 1997 Peter Metz

Table of Contents

Table of Contents of Volume 190
Stereoselective Heterocyclic Synthesis II

Hetero Diels-Alder Reactions in Organic Chemistry

Lutz F. Tietze* and Georg Kettschau

Institut für Organische Chemie der Georg-August Universität, Tammannstr. 2, D-37077 Göttingen, Germany. FAX Int. 49 (0)5 51 39 94 76

The hetero Diels-Alder reaction is one of the most important methods for the synthesis of heterocycles. In this article an overview is given for the period since 1989 describing the reaction of heterobutadienes such as oxabutadienes, thiabutadienes, azabutadienes, diaazabutadienes, nitroso-alkenes and nitroalkenes as well as of heterodienophiles such as carbonyls, thiocarbonyls, imines, iminium salts, azo- and nitroso compounds. In addition, several other less common hetero Diels-Alder reactions such as cycloadditions of thiaazabutadienes, oxa-azabutadienes, dioxabutadienes, dithiabutadienes, oxathiabutadienes, diazaoxabutadienes as well as the use of N-sulfinyl-phosphaalkynes and other dienophiles are mentioned. A main point of discussion is the stereoselectivity of the reactions and the preparation of enantiopure compounds either using dienes and dienophiles carrying a chiral auxiliary or employing chiral Lewis acids. A point stressed is the synthesis of natural products using hetero Diels-Alder reactions leading to carbohydrates, alkaloids, terpenes, antibiotics, mycotoxins, cytochalasans, antitumor agents and several other classes of natural products.

Another topic is the use of high pressure in hetero Diels-Alder reactions discussing the influence on the rate constants and the stereoselectivity. Finally, modern developments such as reactions on solid phase, the use of catalytic monoclonal antibodies, transformations in aqueous solution and the microwave activation are described.

Keywords

Diels-Alder reaction, cycloaddition, azabutadienes, oxabutadienes, diastereoselectivity, enantioselectivity, heterocycles, natural products, high pressure, Lewis acids, dihydropyrans, piperidines, carbohydrates, alkaloids

Contents

List of Abbreviations

Ac	acetyl
AIBN	2,2′-azobisisobutyronitrile
ATP	aluminium tris(2,6-diphenylphenoxide)
9-BBN	9-borabicyclo[3.3.1]nonane
$BF_3 \cdot OEt_2$	boron trifluoride – diethyl ether complex
BINOL	β-binaphthol
Bn	benzyl
Boc	*tert*-butoxycarbonyl
Bz	benzoyl
CAN	ammonium cerium(IV) nitrate

Cbz	benzyloxycarbonyl
CSA	camphorsulfonic acid
DABCO	1,4-diazabicyclo[2.2.2]octane
DAGOH	diacetone glucose
Danishefsky's diene	1-methoxy-3-trimethylsilyloxy-1,3-butadiene
DBN	1,5-diazobicyclo[4.3.0]non-5-ene
DBPO	dibenzoyl peroxide
DBU	1,8-diazabicyclo[5.4.0]undec-7-ene
DCC	dicyclohexylcarbodiimide
DDQ	2,3-dichloro-5,6-dicyano-1,4-benzoquinone
DEAD	diethyl azodicarboxylate
DET	diethyl tartrate
DIBAL-H	diisobutylaluminium hydride
DIPT	diisopropyl tartrate
DMAP	*N,N*-dimethylaminopyridine
DMB	3,4-dimethoxybenzyl
DME	1,2-dimethoxyethane
DMF	dimethylformamide
DMPU	1,3-dimethyl-3,4,5,6-tetrahydro-2(1*H*)pyrimidinone
DMSO	dimethyl sulfoxide
EDDA	ethylene diammonium diacetate
$Eu(dppm)_3$	di(perfluoro-2-propoxypropionyl)methanato europium
$Eu(fod)_3$	6,6,7,7,8,8,8-heptafluoro-2,2-dimethyl-3,5-octanedionato europium
$Eu(hfbc)_3$	3-(heptafluorobutyryl)-(+)-campherato europium
$Eu(hfc)_3$	3-(heptafluoropropylhydroxymethylene)-(+)-campherato europium
Fmoc	9-fluorenylmethoxycarbonyl
HMPA	hexamethylphosphoric triamide
Ipc	isopinocamphenyl
KHMDS	potassium hexamethyldisilazanide
LDA	lithium diisopropylamide
MCPBA	3-chloroperoxybenzoic acid
MEM	(2-methoxyethoxy)methyl
Mes	mesityl
MOM	methoxymethyl
MS	molecular sieves
NBS	*N*-bromosuccinimide
NCS	*N*-chlorosuccinimide
NIS	*N*-iodosuccinimide
(+)-NLE	positive non linear effect
PCC	pyridinium chlorochromate
PDC	pyridinium dichromate
Ph_3P	triphenylphosphane
Pht	phthaloyl
PMB	4-methoxyphenyl
PNB	4-nitrobenzyl

Pv	pivaloyl
py	pyridine
Red-Al	sodium bis(2-methoxyethoxy)aluminium hydride
SEM	2-(trimethylsilyl)ethoxymethyl
TBAF	tetrabutylammonium fluoride
TBDMS	*tert*-butyldimethylsilyl
TBDMSCl	*tert*-butyldimethylsilyl chloride
TBDMSOTf	*tert*-butyldimethylsilyl triflate
TBDPS	*tert*-butyldiphenylsilyl
Tf	trifluoromethanesulfonyl
thexyl	1,1,2-trimethylpropyl
THF	tetrahydrofuran
THP	tetrahydropyran-2-yl
TIPDS	1,1,3,3-tetraisopropyldisiloxane-1,3-diyl
TIPS	triisopropylsilyl
TMEDA	*N,N,N′,N′*-tetramethylethylenediamine
TMS	trimethylsilyl
TMSBr	trimethylsilyl bromide
TMSCl	trimethylsilyl chloride
TMSI	trimethylsilyl iodide
TMSOTf	trimethylsilyl triflate
Tol	4-methylphenyl
Ts	4-toluenesulfonyl

1 Introduction and General Aspects

1.1 Introduction

The Diels-Alder reaction formally describes the addition of a 1,3-diene and a dienophile to give a six-membered ring system with one or two double bonds depending on the type of dienophile used (Fig. 1-1). It is clearly, even today, one of the most important synthetic procedures since its first general description in 1928 by Diels and Alder [1], since it meets the requirements of a modern synthetic method [2, 3, 4] to a great extent by showing an excellent chemo- and regioselectivity as well as a high simple and induced diastereoselectivity in many cases. Furthermore, it possesses a high atom efficiency as well as bond forming efficiency and permits the synthesis of complex molecules from simple starting materials especially in those cases where the diene or dienophile is formed in situ in a domino type transformation [3–6].

Fig. 1-1

Most importantly, the scope of the Diels-Alder reaction is very high – not only allowing the synthesis of cyclohexenes and 1,4-cyclohexadienes using 1,3-butadienes and alkenes and alkynes, respectively, but also giving access to a multitude of different heterocycles by exchanging the atoms **a**–**d** in the butadiene as well as the atoms **e** and **f** in the alkene by heteroatoms such as oxygen, nitrogen and sulfur. However, also dienes and dienophiles with several other atoms as phosphorous, boron, silicone, and selenium have been described. Thus, many different heterodienes and heterodienophiles have been developed over the years (Tables 1-1 and 1-2).

Several reviews and books have already appeared on the hetero Diels-Alder reaction [3–23]. The latest general overlook are the articles of Boger and Weinreb in Comprehensive Organic Synthesis covering the literature until 1989.

In this article we describe novel developments in the synthesis of heterocycles by hetero Diels-Alder reactions covering the literature from 1989. However, as a background and if neccessary for the understanding, also older publications will be presented. Due to the restriction of space only the most important and synthetically most useful dienes and dienophiles which are displayed in Table 1-1 and Table 1-2 will be discussed in this article.

Clearly, an important feature will be the selectivity of these reactions. In this respect, the control of *endo*- and *exo*-selectivity using different Lewis acids, the induced diastereoselectivity with chiral heterobutadienes as well as chiral heterodienophiles and finally the use of chiral Lewis acids for the enantioselective synthesis will be discussed. In recent time some attention has been paid to hetero Diels-Alder reactions in aqueous solutions and in the presence of inor-

Table 1-1. Selected heterodienophiles for the Diels-Alder reaction

O=C< aldehydes, ketones	N=C< imines iminium salts	N=N azo compounds	O=N nitroso compounds
$^{+}S(O^{-})=N$ N-sulfinylimines	S=C< thioaldehydes	S=S diatomic sulfur	O=S=O sulfur dioxide
Se=C< selenoaldehydes	$O=S^{+}(O^{-})=N$ N-sulfonylimines	P≡C– phosphaalkynes	N≡C– nitriles

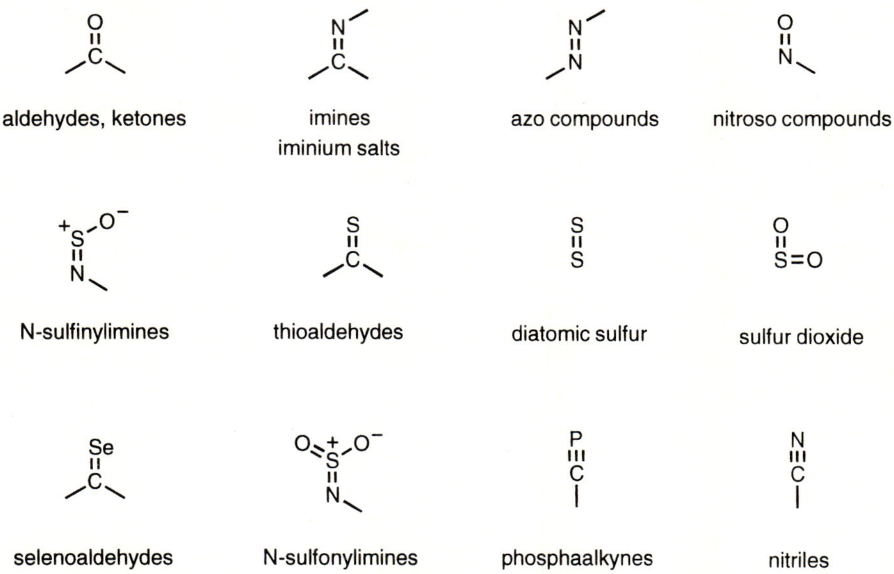

Table 1-2. Selected heterodienes for the Diels-Alder reaction

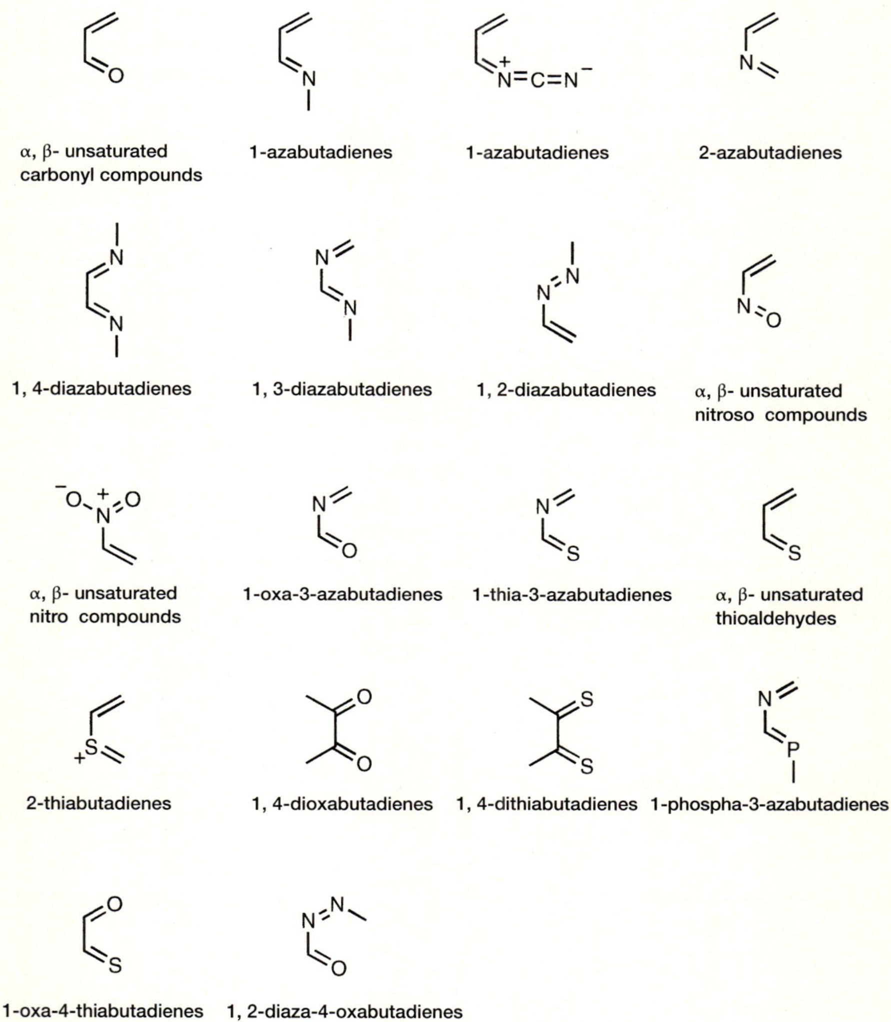

ganic salts. Also sonification and microwave irradiation under solvent-free conditions have been used. In a few cases hetero-Diels-Alder reactions are induced electro- or photochemically. Another interesting point is the use of catalytic antibodies, not only to accelerate the Diels-Alder reactions, but also to allow the synthesis of cycloadducts having a configuration which otherwise could not be obtained.

The reaction rate and the selectivity of hetero Diels-Alder reactions can also be influenced by applying high pressure. A large amount of knowledge has been obtained in this field in the last years with detailed information about $\Delta H^{\#}$, $\Delta S^{\#}$, $\Delta V^{\#}$ and $\Delta\Delta V^{\#}$. Futhermore, the use of hetero Diels-Alder reactions in the synthesis of natural products such as alkaloids, antibiotics, carbohydrates, hetero-steroids, iridoids, macrocycles, mycotoxins and polyethers will be discussed. Finally, a first example of a hetero Diels-Alder reaction on solid support will be given.

The question whether Diels-Alder reactions also occur in nature can not be answered yet since special enzymes catalyzing these reactions have not been found so far [24–25]. However, artificial catalytic antibodies for Diels-Alder reactions are well known [26] and recently an all-carbon [4+2]cycloaddition has been observed in the biosynthesis of two phytotoxic solanopyrones **1-1** and **1-2** from the fungus *Alternaria solani* using a cell-free extract of this organism [27]; it is highly probable that the involved enzymes will soon be isolated (Fig. 1-2).

Another all-carbon Diels-Alder reaction is proposed for the biosynthesis of the indole alkaloids tabersonine **1-6** and catharanthine **1-7** of the Aspidosperma and Iboga family [28–31]. The compounds are formed via strictosidine **1-3**, the first nitrogen-containing precursor of the monoterpenoid indole alkaloids, and stemmadenine **1-4**, which is cleaved to give the proposed intermediate dehydrosecodine **1-5** with an acrylate and a 1,3-butadiene moiety (Scheme 1-1).

A hetero Diels-Alder reaction of a precursor **1-9** may be involved in the biosynthesis of the lignane carpanone **1-8** (Fig. 1-3), however, there is no proof for such an assumption [32]. On the other hand, it is well known that pericyclic reactions such as electrocyclic reactions and sigmatropic rearrangements occur in nature e.g. in the biosynthesis of vitamine D, vitamine B_{12} [33–35] and ectocarpene [36].

In this article a differentiation of concerted and two-step cycloadditions will not be made although this point will be briefly discussed in the theoretical part. However, products which could be formally formed by a hetero Diels-Alder reaction, but for which a different mechanism has been proven will not be included. Thus, quite recently it has been shown that the formation of an oxazine by reaction of *N*-sulfinyl-*p*-toluenesulfonamide and an excess of propanal in the presence of boron trifluoride etherate does not involve a hetero Diels-Alder reaction [37].

1-1 **1-2**

Fig. 1-2

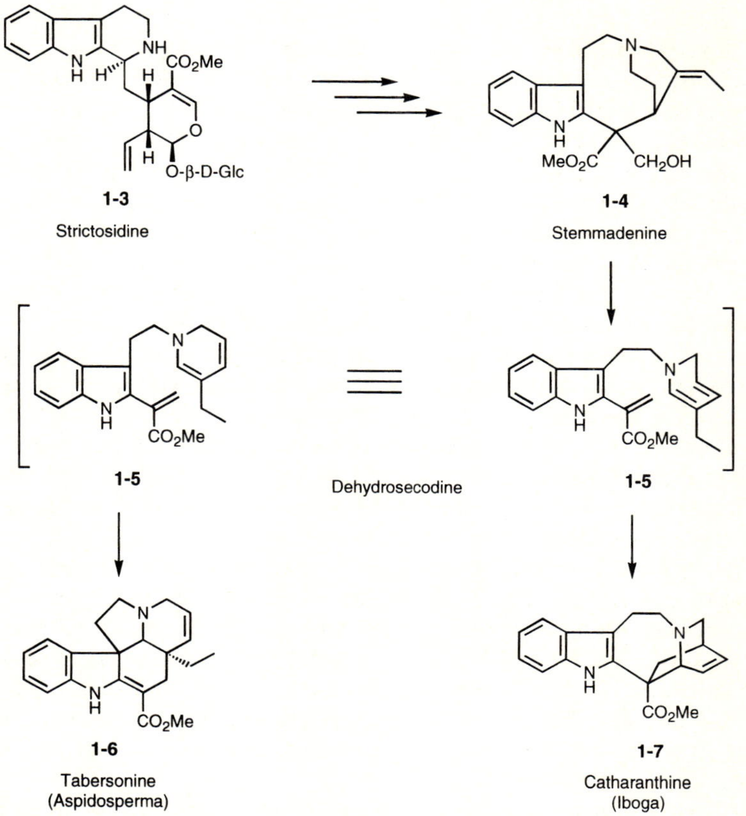

1-3
Strictosidine

1-4
Stemmadenine

1-5

Dehydrosecodine

1-5

1-6
Tabersonine
(Aspidosperma)

1-7
Catharanthine
(Iboga)

Scheme 1-1. Proposed biosynthesis of indole alkaloids via a hetero Diels-Alder reaction

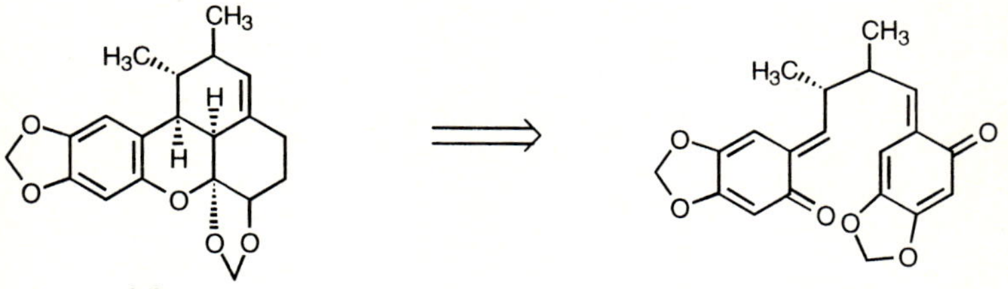

1-8

1-9

Fig. 1-3

1.2
Stereochemical and Theoretical Aspects of Hetero Diels-Alder Reactions

The stereochemistry of the cycloadducts in hetero Diels-Alder as well as of the all-carbon Diels-Alder reactions depends upon the different geometry of the possible transition structures [3, 12, 38]. According to an *endo*- or *exo*-orientation of the dienophile and an (*E*)- or (*Z*)-configuration of the diene, four different transition structures have to be discussed which are shown exemplary for 1-oxa-1,3-butadienes in the inter- and intramolecular mode (Schemes 1-2 and 1-3).

There is no clear definition for the *endo*- and *exo*-orientation for intermolecular reactions. Usually the orientation with the substituent at the dienophile under or above the diene is called *endo*, however, in the case where two or more different substituents at the dienophile exist, this rule cannot be used anymore. We therefore suggest that the following rule should be applied for intermolecular Diels-Alder reactions: *The orientation of the dienophile with the substituent having the highest priority according to the Cahn-Ingold-Prelog rules lying under or above the diene is called endo. The opposite is called exo.* For hetero Diels-Alder reactions a slight modification is necessary: *The orientation of the dienophile with the substituent at the centre being closest to the terminal heteroatom in the diene according to the product, which has the highest priority according to the Cahn-Ingold-Prelog rules lying under or above the diene is called endo. In cases of two terminal heteroatoms the one with the highest priority counts; in cases of no terminal heteroatom, the next heteroatom counts.* For intramolecular Diels-Alder reactions the known definition should be used; thus, *the orientation with the chain connecting the diene and dienophile lying under or above the diene is called endo.*

From the transition structures, it can be seen that the *cis*-product can be formed either by an *endo-E-syn* or an *exo-Z-syn* orientation, whereas the *trans*-product is obtained either by an *exo-E-anti* or an *endo-Z-anti* transition structure. For intramolecular reactions the situation is simplified since calculations have

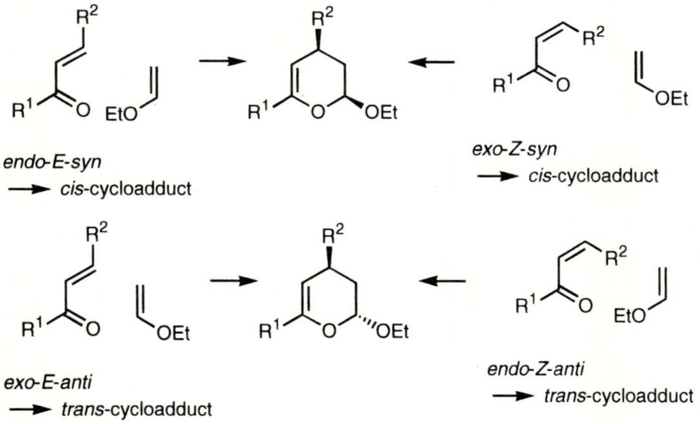

endo-E-syn → *cis*-cycloadduct

exo-Z-syn → *cis*-cycloadduct

exo-E-anti → *trans*-cycloadduct

endo-Z-anti → *trans*-cycloadduct

Scheme 1-2. Transition structures for the intermolecular hetero Diels-Alder reaction of 1-oxa-1,3-butadienes

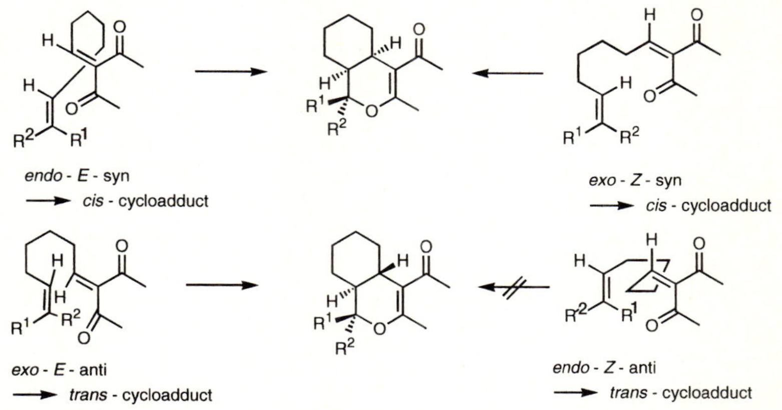

Scheme 1-3. Transition structures for the intramolecular hetero Diels-Alder reaction of 1-oxa-1,3-butadienes

shown that the *endo-Z-anti* transition structure needs not to be considered due to its high energy [54]. However, when discussing the stereochemistry of hetero Diels-Alder reactions and naturally also of the all-carbon cycloadditions one should keep in mind that the configuration of the diene in the ground state does not have to be the configuration of the reacting diene. Especially heterodienes can isomerise quite easily [38]. Thus, reaction of the (*Z*)-1-oxa-1,3-butadiene **1-10** obtained by a Knoevenagel condensation of the corresponding benzaldehyde and the pyrazolone gave nearly exclusively the *cis*-fused product **1-11** at 80 °C via an *endo-E-syn* transition structure (*cis*:*trans* = 17:1). This interpretation is confirmed by the fact that the *tert*-butyl derivative **1-12** does not react at 80 °C due to the lack of isomerisation about the double bond. However, under irradiation which facilitates the isomerisation the cycloaddition takes place also at 80 °C (*cis*:*trans* = 50:1) (Fig. 1-4).

In contrast to the great number of calculations concerning the all-carbon Diels-Alder reaction [39], there are only a few theoretical studies on the hetero Diels-Alder reaction [41, 42, 45–53]. The general mechanism of the Diels-Alder reaction is still in discussion; however, in most cases a concerted reaction is assumed, but there is also evidence for a two-step path. The ab initio calculations carried out for the butadiene/ethene system by Houk, Ortega, Bernardi und Gajewski gave a symmetrical transition structure; only using the semiempirical AM1/CI method (half electron approximation) an unsymmetrical diradicaloid intermediate was found [40].

For hetero Diels-Alder reactions it has been shown by calculations that the transition structures are usually less symmetric than for the all-carbon Diels-Alder reactions; also a change from a concerted non-synchronous to a stepwise mechanism depending on the substituents at the reacting species and the reaction conditions can occur.

One of the first calculations on hetero Diels-Alder reactions was done in our group in collaboration with Anders on the 1-oxa-1,3-butadiene (acrolein)/ethene system [41, 42]. The performed ab initio und semiempirical calculations show

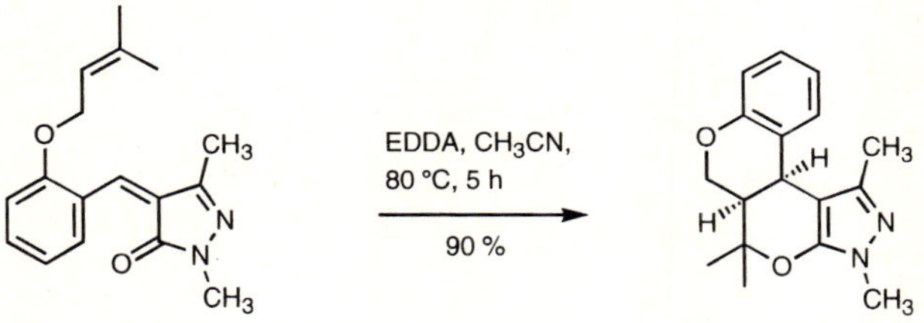

Fig. 1-4

that two competing reaction channels exist, a concerted and a two-step path. All methods used reveal a preference for a one-step mechanism, which is in agreement with experimental observations.

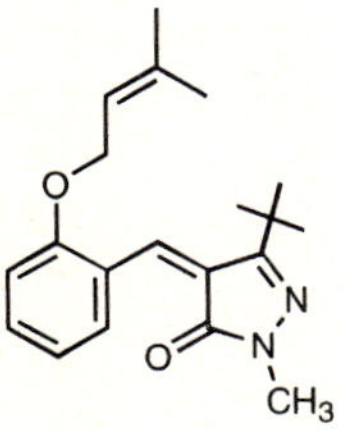

In contrast, for the 1-aza-1,3-butadiene/ethene system the ab initio and semiempirical calculations show a preference for a two-step mechanism [43] which again is in agreement with experimental observations [44, 45].

Houk [46, 47] as well as Jursic and their groups [48] have investigated the hetero Diels-Alder reaction of 1,3-butadienes with heterodienophiles such as formaldehyde, thioformaldehyde, formaldimine, *N*-methylformaldimine, diazene, nitrosyl hydride, singlet oxygen and some BH_3-coordinated and protonated species. Asynchronous transition structures were located with asymmetric heterodienophiles whereas with symmetrical dienophiles a synchronous transition structure was produced. Importantly, the transition structures with *exo* oxygen or nitrogen lone pairs have lower energies than the corresponding *endo* lone pair transition structures.

Further calculations were performed on 2,3-diaza-1,3-butadiene with different heterodienophiles such as ethene, formaldehyde and formaldimine showing the same *exo* oxygen or nitrogen lone pair preference [49] as well as on the nitrosoethene/ethene system [50]. Recently, ab initio studies have also been performed for the Lewis acid catalysed hetero Diels-Alder reaction of isoprene and sulfur dioxide by Sordo [51].

In addition to the purely mechanistical studies calculations on the stereochemistry of more complex molecules have been performed; the data obtained nicely matched the experimental results [54]. Finally, several experiments have been performed to prove the concertedness of the hetero Diels-Alder reactions of 1-oxa-1,3-butadienes [55] and show that the transition structure is unsymmetrical [56].

2
Oxa Diels-Alder Reactions

2.1
Oxa Diels-Alder Reactions with C=O Dienophiles

The [4 + 2]cycloaddition of the carbonyl group of aldehydes as well as of ketones and 1,3-butadienes is a well established method for the synthesis of 5,6-dihydropyrans which are useful substrates for the preparation of carbohydrates and many other natural products. Several excellent reviews on this topic have appeared [10–12, 14, 22]. The first example of this type of reaction using 2,4-dimethyl-1,3-butadiene and formaldehyde to give the 2,4-dimethyl-5,6-dihydro-2*H*-pyran in 60% yield was published by Gresham and Steadmen in 1949 (Scheme 2-1, Eq. 1) [57].

However, employing higher aldehydes the yield was very poor, whereas with chloral reasonable results could again be obtained [58].

Thus, the scope of the cycloaddition is a little limited since only electron-deficient carbonyl groups as in chloral, glyoxylate, oxomalonate, 1,2,3-triketones as well as similar compounds and butadienes with electron-donating groups give high yields. Some older examples are shown in scheme 2-1 [57–63]. By using Lewis acids and on the other hand, applying high pressure, good results have also been seen with less reactive substrates in many cases. In addition, the use of chiral Lewis acids allows an enantioselective cycloaddition. In the synthesis of

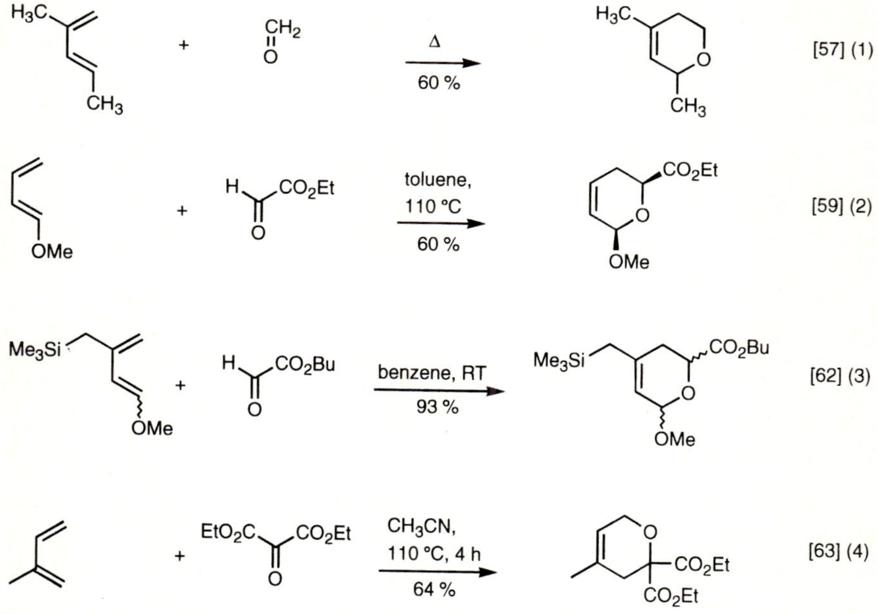

H_3C ... + $CH_2{=}O$ → (Δ, 60 %) ... [57] (1)

... + $H(C{=}O)CO_2Et$ → (toluene, 110 °C, 60 %) ... [59] (2)

Me_3Si ... OMe + $H(C{=}O)CO_2Bu$ → (benzene, RT, 93 %) ... [62] (3)

... + $EtO_2C(C{=}O)CO_2Et$ → (CH_3CN, 110 °C, 4 h, 64 %) ... [63] (4)

Scheme 2-1 Older work of hetero Diels-Alder reactions of carbonyl groups

the perfumery product (+)-ambrenolide **2-1** neither the thermal nor the Lewis acid catalyzed cycloaddition was successful. However, using high pressure the desired product could be obtained, though in low yield Fig. 2-1 [64].

The reactions usually proceed with retention of the configuration of the diene moiety and high regioselectivity, which is controlled by the coefficients of the LUMO of the carbonyl group and of the HOMO of the diene; however thermally, the *endo/exo*-selectivity is low giving normally only a slight excess of the *cis*-compounds (e.g. *cis*:*trans* $\approx$ 2:1, Scheme 2-1, Eq. 2). At higher temperature (e.g. 150 °C) or in the presence of an acid isomerisation can take place to give predominantly the *trans*-compounds (*trans*: *cis* = 4:1) [59].

In a thermal reaction hexafluoroacetone [65] and several trifluoroacetones [66] also react with butadienes in a straightforward fashion.

The advantagous use of Lewis acids in the hetero Diels-Alder reaction of carbonyl compounds has intensively been studied and employed by Danishefsky et al. [23, 67-70]. They also showed that rare earth cations are excellent and mild catalysts due to their high oxophilicity, even allowing the isolation of the highly sensitive primary cycloadducts.

The combination of two all-carbon and one hetero Diels-Alder reaction in the presence of catalytic amounts of $Eu(fod)_3$ of **2-2** and **2-3** to give **2-4** was used in the synthesis of vincomycinone B_2 methyl ester **2-5,** Scheme 2-2 [71].

A recent general study on the reactivity of 3-mono-*O*-activated dienes **2-6** having an alkyl group at C-1 and **2-7** in the presence of a Lewis acid was performed by Palenzuela et al. [72]. The best yields of the cycloadducts **2-8** were obtained with $BF_3 \cdot OEt_2$ in diethyl ether with an *endo/exo*-selectivity of 6:1 (Fig. 2-2). Good results were also found with $LiBF_4$ in acetonitrile/benzene. Aldol reactions [73], silatropic ene reactions [74] and loss of the silyl group [75] were not observed under these conditions.

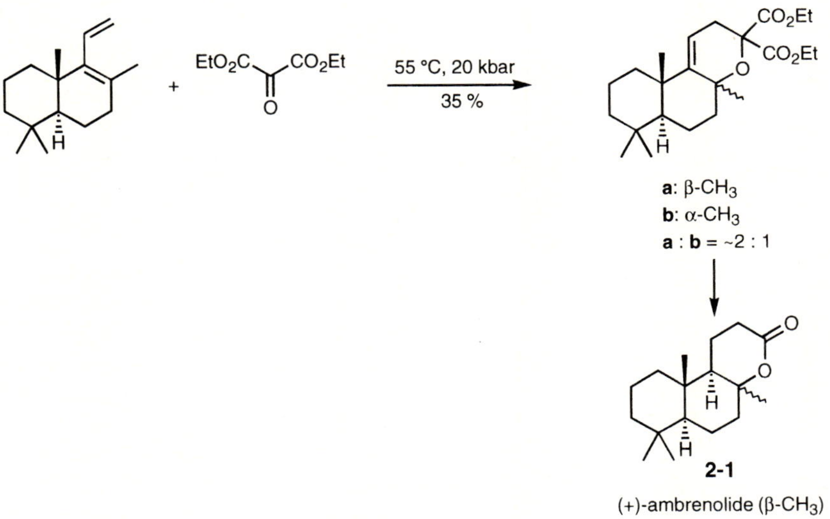

Fig. 2-1

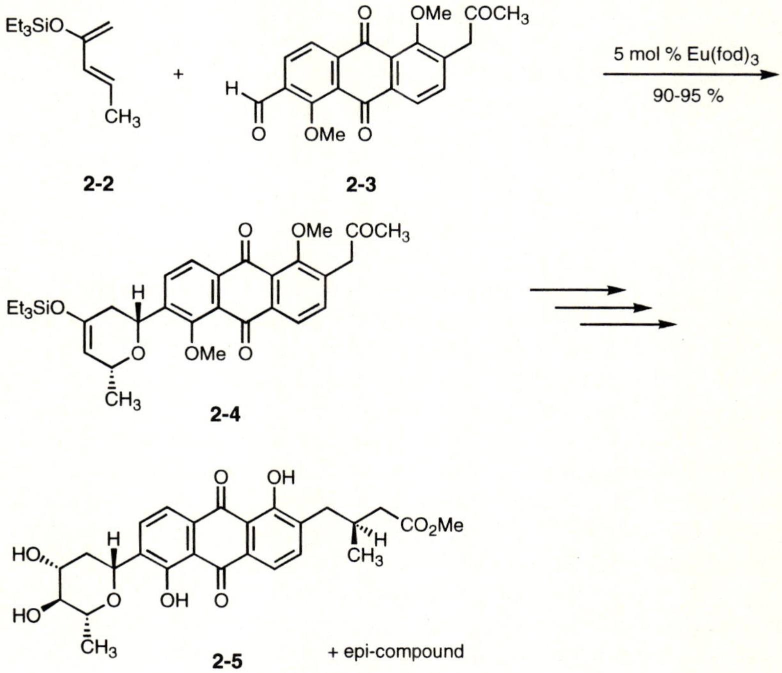

Scheme 2-2

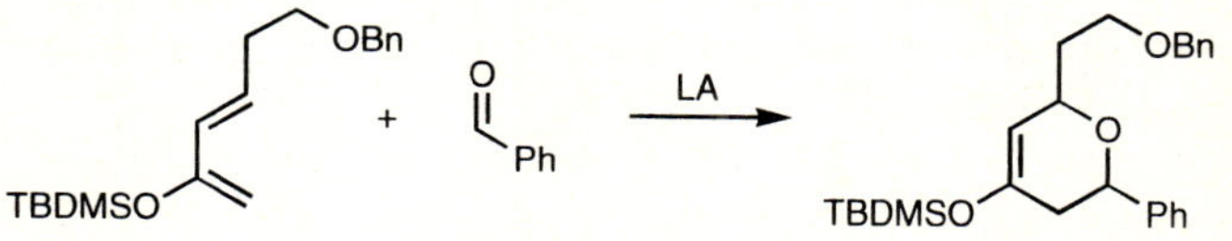

2-6 2-7 2-8

$BF_3 \cdot OEt_2$, Et_2O, -78 °C, < 5 min : 86 % (*cis* : *trans* = 6 : 1)
$LiBF_4$, CH_3CN/benzene, 20 °C, 20 min : 84 % (*cis* : *trans* = 6 : 1)

Fig. 2-2

Hoffmann and his group [76] have used the cycloaddition of 4-formylfuran-2(5*H*)-ones **2-9** and butadienes such as 1-methoxy-3-trimethylsilyloxy-1,3-butadiene (Danishefsky's diene) **2-10**, 1-trimethylsilyoxy-1,3-butadiene and 2,4-dimethyl-1,3-butadienes for the construction of a manoalide substructure **2-11** which belongs to a class of nonsteroidal anti-inflammatory agents. According to the substitution of the butadiene different Lewis acids such as the mild $Eu(fod)_3$, the more reactive $AlMe_3/AlCl_3$ and the highly reactive $TiCl_4$ had to be used (Fig. 2-3).

Page et al. [77] have shown that 2,2-disubstituted 2,3-dihydropyrans are obtained in good yield (69-86%) by reaction of electron-deficient ketones **2-13**

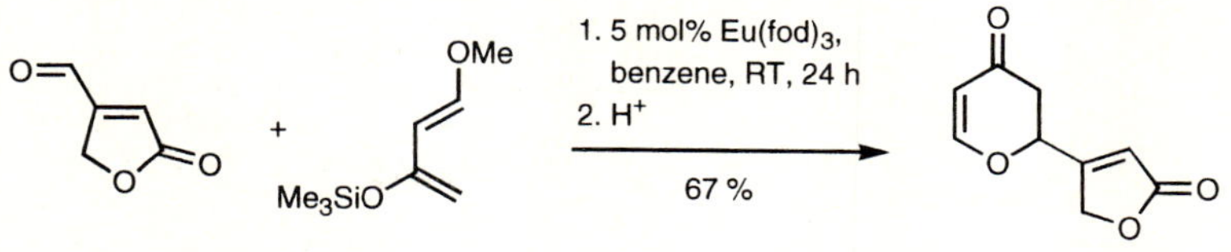

O= ... O; OMe; + Me3SiO; 1. 5 mol% Eu(fod)3, benzene, RT, 24 h; 2. H+; 67 %

2-9 2-10 2-11

Fig. 2-3

with electron-rich dienes **2-12** in the presence of a Lewis acid. The successfully employed substrate types include 1,2-diketones, pyruvates **2-13a**, acylnitriles **2-13b** and oxomalonates **2-13c**. Less electron-deficient carbonyl compounds gave poor results. The choice and quantity of Lewis acid used was of vital importance for these transformations. Neither titanium tetrachloride nor tin tetrachloride proved to be suitable. However, zinc(II) chloride in dry benzene was an effective mediator. The use of substoichiometric amounts gave only low yields. Interestingly, in the case of butane-2,3-dione, no reaction was observed when 2.5 equivalents of the mediator were employed although use of 1.2 equivalents gave 69% yield of the cycloadduct (Fig. 2-4).

A good mediator for the hetero Diels-Alder reaction of aldehydes is the bulky, oxygenophilic aluminium tris(2,6-diphenylphenoxide) (ATP) **2-19** developed by Yamamoto [78] which allows the differentiation between two sterically discriminated aldehydes. Thus, reaction of a mixture of **2-10, 2-16** and **2-17** in the presence of **2-19** gave nearly exclusively **2-15,** whereas in the presence of $BF_3 \cdot OEt_2$ a 1.3:1 mixture of **2-15** and **2-18** was formed (Fig. 2-5).

Recently, also diiodosamarium has been used as catalyst for the hetero Diels-Alder reaction, however the yields and the regioselectivity in these transformations are similar to those with $Eu(fod)_3$ [79].

Good results were also obtained with lithium perchlorate in dichloromethane and diethyl ether. It has been shown that the lithium cation acts as a Lewis acid and the effects are not due to an "internal pressure" [80]. The acceleration is much more pronounced for hetero Diels-Alder reactions as compared to the all-carbon cycloadditions. With chiral aldehydes a high level of chelation control has been observed (see later) [81,82].

It should be noted that the hazardous $LiClO_4$ may be replaced by the less dangerous $LiNTf_2$ [83,84].

A combination of a hetero Diels-Alder reaction of an aldehyde and a radical reaction in a sequential transformation to give the bridged pyrans **2-22** via **2-21** starting from **2-20** containing a seleno moiety and an electron-rich butadiene **2-10,** was described by Clive (Fig. 2-6) [85].

All investigations on the use of Lewis acids in the hetero Diels-Alder reaction of carbonyl compounds clearly show that a careful adjustment of the Lewis acidity for a given system is neccessary. Especially with trimethylsilyloxybutadienes a Mukayama type aldol reaction can easily take place instead of the desired hetero Diels-Alder reaction.

In addition to the use of Lewis acids two further major aspects for this type of cycloadditions have been focused on in recent years. The first important

a : R^1 = Me $R^2 = CO_2Et$ 72 %
b : R^1 = Me R^2 = CN 78 %
c : $R^1 = CO_2Et$ $R^2 = CO_2Et$ 88 %

Fig. 2-4

Fig. 2-5

a : n = 1 **2-21** **a** : 57 % **2-22** **a** : 88 %
b : n = 2 **2-21** **b** : 83 % **2-22** **b** : 99 %

Fig. 2-6

aspect is the application of chiral carbonyl compounds as well as of chiral butadienes either having one or more stereogenic centers in the substrate or bearing a chiral auxiliary. The second aspect concerns enantioselective reactions with chiral mediators or catalysts.

For the diastereoselective hetero Diels-Alder reaction of carbonyl compounds using removable chiral auxiliaries, intensive studies of the uncatalyzed cycloaddition of glyoxylic acid esters with the optically active alcohols menthol, borneol, 2-octanol and 2,2-dimethyl-3-heptanol have been performed. Disappointingly, the induced diastereoselectivities were rather low (0.4–13%) and also applying high pressure did not improve these results to a reasonable extent [86]. In a special case the obtained diastereomers could be separated allowing the entry to thromboxane type derivatives [87].

However, excellent simple (*exo*:*endo*=95:5) and induced diastereoselectivity (94:6) was obtained by Jurczak [88] by applying the bornane sulfone amide derivative of glyoxylic acid **2-23** in the presence of a catalytic amount of a europium salt. Reaction of **2-23** with 1-methoxy-1,3-butadiene **2-24** gave predominantly **2-25a** which was transformed into the lactone **2-26** aiming towards the synthesis of compactin (Fig. 2-7)[89].

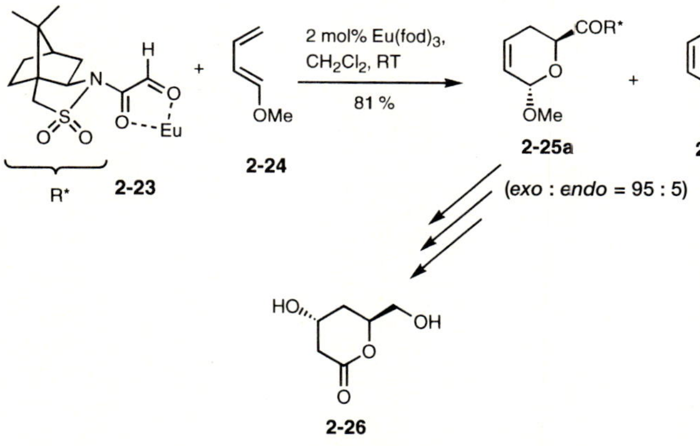

Fig. 2-7

A good induced diastereoselectivity was also found in the cycloaddition of glyoxylates **2-27** to butadienes connected to sugar derivatives such as diacetoneglucose and derivatives of galactose **2-28** at the 3-position as well as of tetrabenzylglucose at different positions even in an uncatalyzed fashion ranging from 73:27 to 96.8:3.2 as shown by David et al. [90, 91]; however, the *endo*/*exo*-selectivity was rather low (Fig. 2-8).

The procedure has been employed for the synthesis of the determinant trisaccharide unit of the human blood group A using **2-28** and the (–)-menthyl glyoxylate **2-27** as a matched pair to give the desired disaccharide **2-30** after isomerisation of the primarily obtained mixture of **2-29** and **2-30**.

An excellent simple and induced selectivity could be obtained by Mulzer and his group [92] in the cycloaddition of 2-trimethylsilyl-oxy-1,3-pentadiene **2-31** and (1*R*,2*S*,5*R*)-8-phenylmenthyl glyoxylate **2-32** [93] in the presence of 0.2 equivalents of anhydrous $MgBr_2$ in THF at 0 °C. After acidic workup the ketone **2-33** was isolated as a single diastereomer (>98%), which was then used for synthesis of the C-26-C-32 tetrahydropyran moiety of swinholide. In contrast,

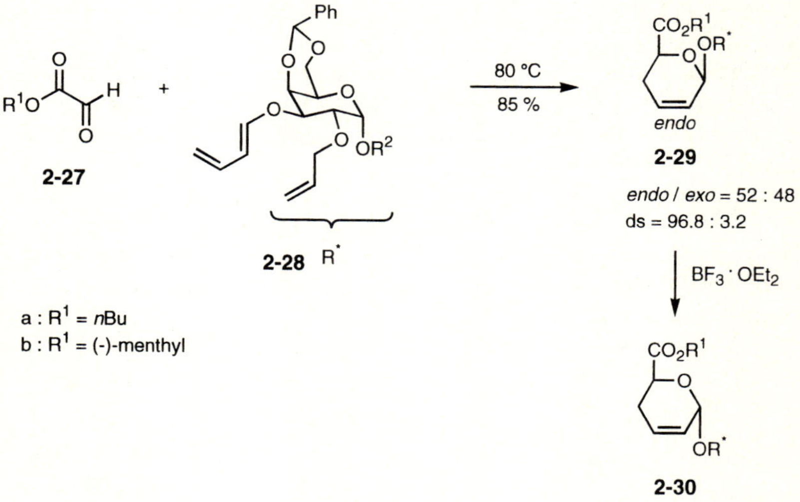

Fig. 2-8

reaction of the (1*R*, 2*S*, 5*R*)-menthyl glyoxylate led to a 1:1-diastereomeric mixture with respect to the auxiliary (Fig. 2-9).

The use of $MgBr_2$ is of importance since the thermal reaction of **2-32b** led to a 2:1-mixture of *rac*-**2-33b** and its *cis*-isomer, whereas with $MgBr_2$ rac-**2-33b** was obtained in an exo/endo-selectivity of >95:5. The *endo*-transition structure of the intermediately formed Mg-chelate of **2-32b** and also of **2-32a** is disfavoured since steric interactions between the bromine and the carbon skeleton of the diene occur in this arrangement.

Recently, Breitmaier et al. [94] showed that in the hetero Diels-Alder reaction of triketones such as indantrione **2-34** and alloxane with the chiral 2-methyl-1-(1-phenylalkoxy)-1,3-butadienes **2-35** a good diastereoselectivity can be obtained. The cycloaddition proceeded regioselectively with increasing facial selectivity in correlation to the steric demand of the alkyl group at the benzylic position in the auxiliary to give the dihydropyran **2-36** as the major and **2-37** as the minor product (Fig. 2-10).

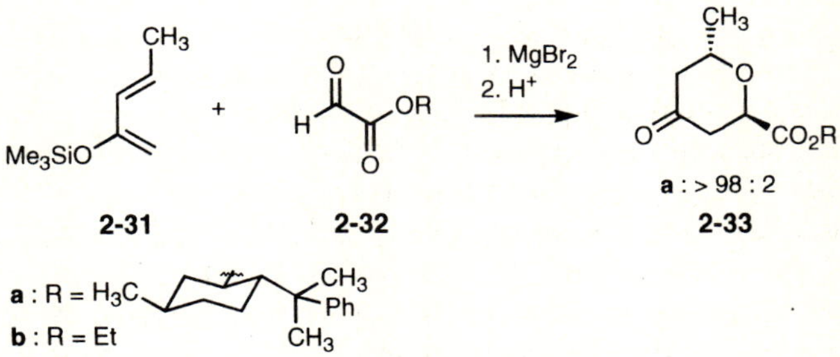

Fig. 2-9

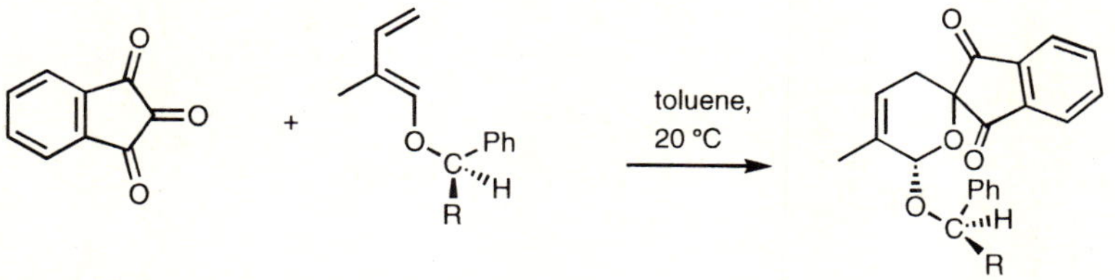

toluene, 20 °C

2-34 + **2-35** → **2-36** major diastereomer + **2-37** minor diastereomer

Diasteroselectivity

	R	**2-36** : **2-37**
a	Me	78 : 22
b	Et	84 : 16
c	*i*Pr	88 : 12
d	*t*Bu	94 : 6

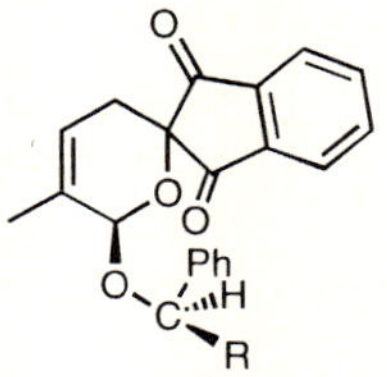

Fig. 2-10

Excellent diastereoselectivities were also obtained in the cycloadditions of chiral 3-(*p*-tolylsulfinyl)-2-furaldehyde **2-38** and 1-methoxy-3-(trimethylsilyloxy)-1,3-butadiene **2-10** in the presence of lanthanoid Lewis acids as described by Arai [95]. Noteworthy, the reaction of **2-38** and **2-10** in the presence of $Yb(OTf)_3$, $Nd(OTf)_3$ or $Sm(OTf)_3$ provided the cycloadduct **2-39** as the major diastereomer, whereas with $Eu(thd)_3$ the corresponding diastereomer **2-40** was obtained predominantly. The use of $ZnCl_2$ as Lewis acid provided the product **2-39** and **2-40** as a 1: 1-mixture (Fig. 2-11).

Chiral aldehydes such as *N*-protected α-aminoaldehydes and α-alkoxyaldehydes as well as chiral butadienes derived from sugars by a Wittig reaction have also been used in the hetero Diels-Alder reactions successfully with the inducing stereogenic centers remaining in the obtained cycloadducts.
The reaction of *N*-(*tert*-butoxycarbonyl)leucinal **2-41 a** by Danishefsky et al. with 1-methoxy-3-trimethylsilyloxy-1,3-butadiene **2-10** gave the pyrones **2-42** and **2-43** with an induced diastereoselectivity of 9:1 in favour of the *syn*-compound in the presence of $Eu(hfc)_3$ [96]. Later Garner [97] used a *N*-Boc-serine derived aldehyde **2-41 b** and Danishefsky's diene **2-10.** In both cases a chelation-control forming a complex between the nitrogen and the oxygen could explain the obtained selectivity. In the presence of HMPA chelation is minimized to give a higher extent of the *anti*-product **2-43** (Fig. 2-12) [97].

In a similar way Midland [98, 99] investigated the reaction of 1,3-dimethoxy-1-(trimethylsilyoxy)butadiene **2-45** [100] with a variety of *N*-protected α-aminoaldehydes e.g. **2-44 a–c** in the presence of several Lewis acids as $Eu(hfc)_3$ and Et_2AlCl, of which the latter gave the best selectivities. Using **2-44 a** and **2-45** in the presence of $Eu(hfc)_3$ the pyrones **2-46 a** and **2-47 a** were obtained in a ratio of

	2-39		2-40
$Yb(OTf)_3$	96.5	:	3.5
$Nd(OTf)_3$	99	:	1
$Sm(OTf)_3$	98.5	:	1.5
$Eu(thd)_3$	11.5	:	88.5

Fig. 2-11

a : CH_2-$CH(CH_3)_2$
b : CH_2OH

Fig. 2-12

80:20, whereas upon reaction of **2-44b** and **2-45** in the presence of Et_2AlCl a ratio of 92:8 of **2-46b** and **2-47b** was found. Interestingly, with the *N*,*N*-dibenzyl derivative **2-44c** [101] the selectivity was completely reversed to give **2-46c** and **2-47c** in a ratio of 1:99. The results are consistent with a chelation control in the former reaction, whereas **2-44c** reacts in a Cram-type fashion (Fig. 2-13). In a similar way, also protected α-hydroxyaldehydes were used.

a : R^1 = Me, R^2 = Cbz, R^3 = H
b : R^1 = *i*Pr, R^2 = Boc, R^3 = H
c : R^1 = Me, R^2 = Bn, R^3 = Bn

Fig. 2-13

As already mentioned, also $LiClO_4$ can be employed to obtain a high level of chelation control. Thus, Reetz and his group [81] observed an excellent induced diastereoselectivity in the hetero Diels-Alder reaction of a chiral α-benzyloxy-propanal **2-48** with the butadiene **2-10** in the presence of 15 mol % $LiClO_4$ in dichloromethane to afford the dihydropyrans **2-49** and **2-50.** A reduction of the amount of catalyst led to a decrease in selectivity. Also $MgBr_2$ could be utilised in this transformation with good results, however, only when used in stoichiometric amounts (Fig. 2-14).

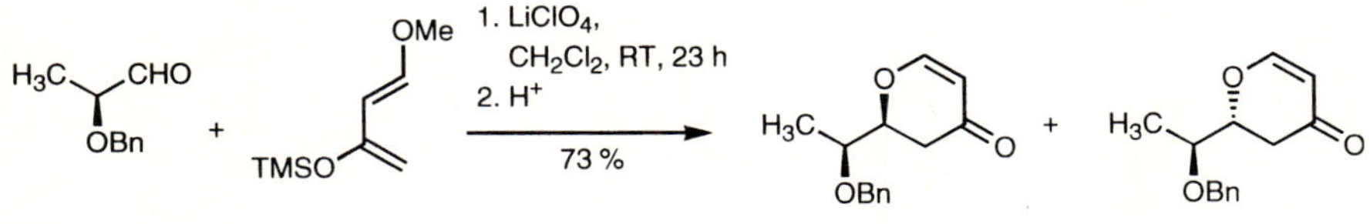

2-48 **2-10** **2-49** **2-50**

	2-49		2-50
6 mol% $LiClO_4$	90	:	10
15 mol% $LiClO_4$	> 95	:	5

Fig. 2-14

At the same time, Grieco et al. [82] have investigated the $LiClO_4$ catalysis for the cycloaddition of *N*-protected α-aminoaldehydes to butadienes such as 1-methoxy-3-trimethylsilyloxy-1,3-butadiene in diethyl ether to give the *syn*-cycloadducts. As already described by Midland [98, 99] for this type of transformation, the diastereofacial selectivity could be reversed by changing the nature of the protecting group on the nitrogen and utilising 3.0 M $LiClO_4$ in diethyl ether. Thus, hetero Diels-Alder reaction of **2-51** and **2-52** under these conditions followed by acidic workup furnished the *anti*-cycloadduct **2-53** as a single diastereomer (Fig. 2-15).

Cycloaddition of the diene **2-54** obtained from 2,4-benzylidene-erythrose by a Wittig reaction with sodium glyoxylate **2-55** in water for 2.5 days at reflux provided a mixture of four adducts in good yield, but in low diastereoselectivity. Interestingly, for the reaction of methyl glyoxylate with the acetylated diene in an organic solvent higher temperature was needed (4 h, 140 °C) with decreased yield (25%) (Fig. 2-16) [102].

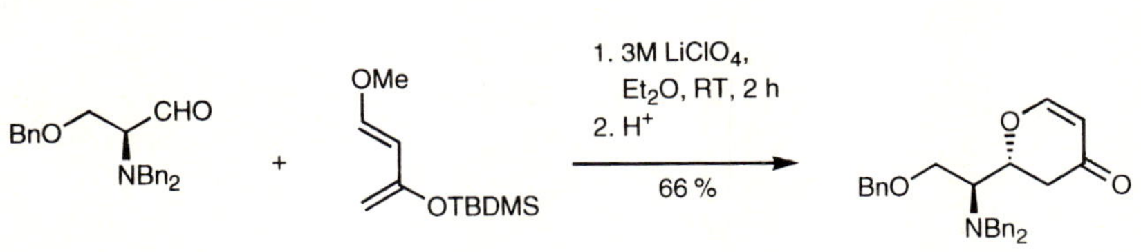

2-51 **2-52** **2-53**

Fig. 2-15

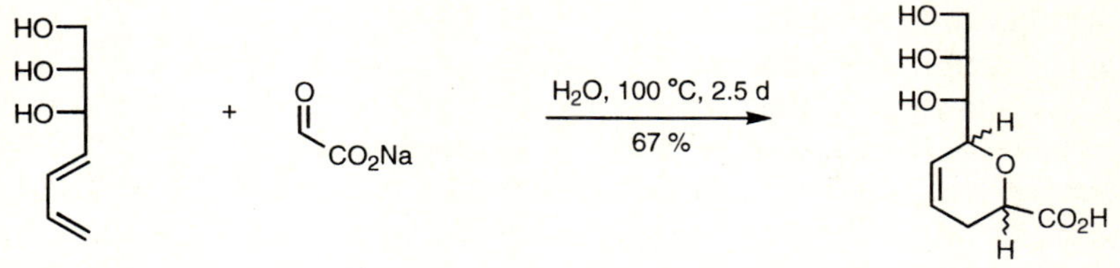

2-54 2-55 2-56

four diastereomers
(28 : 31 : 16 : 25)

Fig. 2-16

Based on the work of Danishefsky [70] on the cycloaddition of aldehydes to dienes bearing various menthyl auxiliaries, Stoodley et al. have used butadienyl glycosides **2-58** with the aim to synthesise (1→1)-linked disaccharides [103]. The reaction was performed with *p*-nitrobenzaldehyde **2-57** in the presence of an europium salt since benzaldehyde itself did not react. Reasonable induction was obtained in the presence of the chiral (–)-Eu(hfc)$_3$ as a matched pair to give **2-59** as the major product (Fig. 2-17).

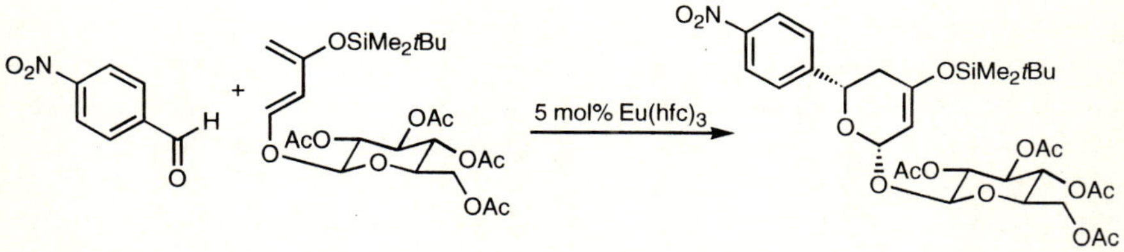

2-57 2-58 2-59

major product : 39 %
(8 : 1 : 1)

Fig. 2-17

In recent years a great improvement in the enantioselective cycloaddition of aldehydes as well as ketones to electron-rich dienes such as **2-10** has been achieved. Danishefsky et al. [68] have used chiral lanthanide complexes of the type Eu(hfbc)$_3$ (hfbc = 3-heptafluorobutyrylcamphor) known as a chiral paramagnetic NMR shift reagent. However, the enantioselectivity using this catalyst was not very high. Similar, less satisfying results were obtained with a cationic Ru complex containing chiral chelating diphosphines [104].

For the first time good enantioselectivities were found by Yamamoto [105, 106] using the chiral organoaluminium reagent **2-63** which, however, is rather difficult to obtain. Reaction of benzaldehyde and the electron rich butadiene **2-60** in toluene at –20 °C for 2 h in the presence of catalytic amounts of the aluminium complex (*R*)-**2-63a** (10 mol%) followed by acidic work up resulted in the formation of a 92:8 mixture of the *cis*-dihydropyrone **2-61** with an *ee* of 95% and the *trans*-product **2-62**. By using the even more hindered aluminium reagent (*R*)-**2-63b** *cis*/*trans*- and enantioselectivity could be improved (Fig. 2-18).

Several different butadienes and aliphatic aldehydes were used with good success. An interesting approach for this transformation is the in situ complexation of one enantiomer of the aluminium complex employing chiral ketones and thus allowing the remaining enantiomer to be utilized as a Lewis acid for the asymmetric synthesis.

A better accessible chiral mediator is the (acyloxy)borane (CAB) **2-64** prepared in situ from a tartaric acid derivative and arylboronic acid at room temperature. Hetero Diels-Alder reaction of benzaldehyde and Danishefsky's diene **2-10** in the presence of **2-64** gave the corresponding pyrone after acidic work up with 52–95% *ee* depending on R. The best results were obtained with R = 2,4,6-Me_3Ph and 2,4,6-iPr_3Ph. Similarly, with **2-60** the pyrone **2-61** with up to 97% *ee* was found [107].

The outstanding properties of binaphthol (BINOL) as a ligand in chiral Lewis acidic metal complexes were also demonstrated highly successfully by Mikami [108, 109] using a binol-titanium complex **2-69a**. Even in the cycloaddition of methyl glyoxylate **2-66** to 1-methoxy-1,3-butadiene **2-65** which usually shows only a low selectivity, a reasonable *cis*/*trans*-selectivity and an excellent enantioselectivity could be obtained in the presence of catalytic amounts of this complex.

Noteworthy, in contrast to earlier work with the complex **2-69a** in the presence of molecular sieves (MS), the MS-free system gave a better *endo*-selectivity and enantioselectivity. Interestingly, a positive non linear effect [(+)-NLE] [110] is observed using e.g. a mixture of (*R*)-**2-69a** and (*R*)/(*S*)-**2-69a**; this effect was not found in the presence of a non-racemic mixture of (*R*)-**2-69a** and (*S*)-**2-69a** in

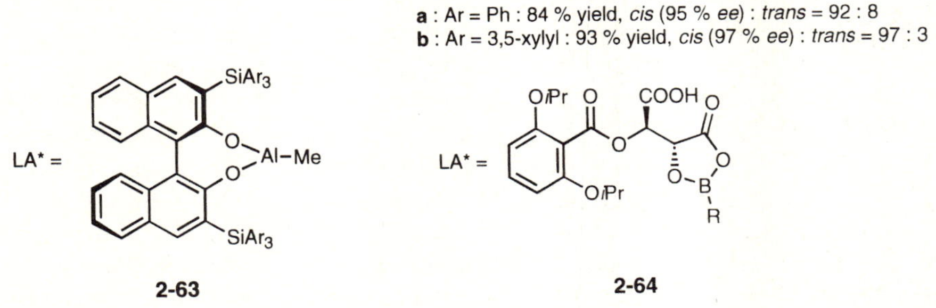

Fig. 2-18

absence of molecular sieves indicating that the catalytically non active (*R*)/(*S*)-dimer-**2-69 a** is not or only slowly formed in absence of molecular sieves.

It should be noted that the new binaphthol catalyst **2-69 b** may give better results in some cases, but not always. Thus, the *cis*-selectivity is decreased in the reaction of **2-65** and **2-66** with nearly identical enantioselectivity (Fig. 2-19).

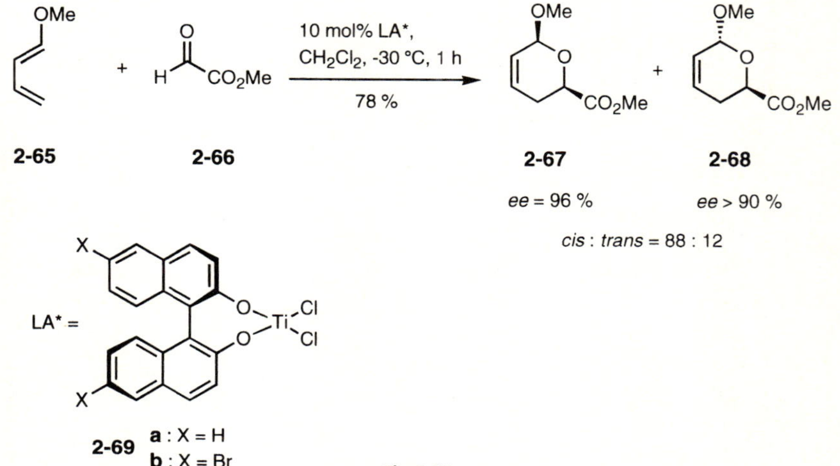

Fig. 2-19

Similarly, Keck [111] has used the Ti(*O*-*i*-Pr)$_4$/BINOL complex (10 mol %) for the hetero Diels-Alder reaction of 1-methoxy-3-trimethylsilyloxy-1,3-butadienes **2-10** and non-activated aldehydes. The lowest enantioselectivity was obtained with benzaldehyde and the best with phenylacetaldehyde and some aliphatic aldehydes to give the corresponding dihydropyrans with *ee* values ranging from 75% up to 97%.

Good enantioselectivities were also found by Togni [112] using a novel optically active oxovanadium(IV) complex **2-72** bearing camphor-derived 1,3-diketonato ligands. The reaction of benzaldehyde with 1-methoxy-2,4-dimethyl-3-(triethylsilyloxy)-1,3-butadiene **2-70** in the presence of 5 mol % of bis[3-(heptafluorobutylryl)camphorato]oxovanadium **2-72** at –78 °C gave the pyrone after acidic work up with a simple diastereoselectivity of 98.5% *de* and 85% *ee*. This was clearly the best result, other examples were less satisfying (Fig. 2-20).

Also good results were obtained by Jørgensen [113] using a chiral copper(II) complex **2-77**. However, employing butadienes bearing a methyl group such as 2,3-dimethyl-1,3-butadiene **2-73** with alkyl glyoxylate **2-74** a mixture of Diels-Alder and ene product was obtained. The observed *ee*-values for both products vary only slightly. Thus, reaction of **2-73** and **2-74** in the presence of **2-77** gave **2-75** with 85% *ee* in 20% yield and **2-76** with 83% *ee* in 36% yield (Fig. 2-21).

It should be noted that the use of polar solvents such as nitromethane leads to a significant improvement of the catalytic properties of **2-77** probably due to

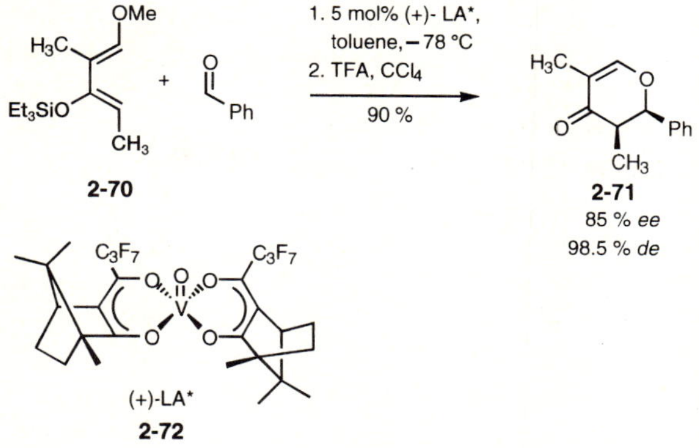

Fig. 2-20

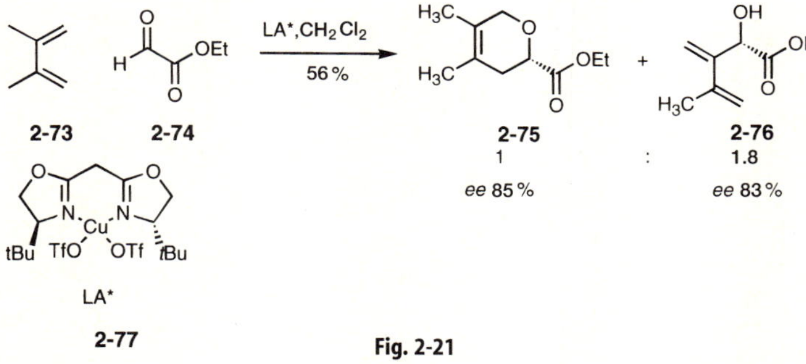

Fig. 2-21

an accelerating effect of ligand dissociation from the metal to give the cationic copper-Lewis acid [114].

A clear two step formation of a pyrone by an enantioselective Mukaiyama-aldol and acid catalysed aldol dihydropyrone annulation using aliphatic and aromatic aldehydes and 1-methoxy-3-trimethylsilyloxy-1,3-butadiene in the presence of a tryptophan-derived oxazaborolidine was described by Corey et al. [115]. The resulting pyrone which could be assigned as a formal Diels-Alder adduct was obtained with a 67–82% *ee* and 57–100% yield.

The observation by Corey again raises the question under which conditions a hetero Diels-Alder or a two-step aldol reaction takes place especially when using silyloxybutadienes. Thus, in several studies a clear structure determination of the intermediate cycloadducts was not performed, being directly transformed into the final pyrone by an acid-catalysed reaction. Under these conditions also a primarily formed aldol-adduct would yield the isolated pyrones.

2.2
Oxa Diels-Alder Reactions with 1-Oxa-1,3-butadienes

The hetero Diels-Alder reaction of α,β-unsaturated aldehydes and ketones with electron rich alkenes such as enol ethers, thioenol ethers, ketene acetals, enamines, alkynyl ethers, ketene aminals and ynamines as well as selected simple alkenes gives an excellent access to 2-substituted 3,4-dihydro-2*H*-pyrans which are useful precursors for the synthesis of carbohydrates, iridoids, mycotoxins and other natural products. Several excellent reviews on this topic have already been published which cover the literature until 1989 [10–12, 14, 22]. The reaction is controlled by a dominant interaction of the LUMO of the 1-oxa-1,3-butadiene and the HOMO of the dienophile and thus belongs to the Diels-Alder reactions with inverse electron demand. It is usually a concerted non-synchronous transformation [41, 42, 54] with retention of the configuration of the dienophile [55]. Electron-withdrawing groups at the 1-oxa-1,3-butadiene greatly enhance their reactivity by substantially lowering the energy of the LUMO of the oxabutadiene allowing the performance of the Diels-Alder reaction without the additon of a catalyst in several cases already at 20 °C even with simple alkenes [12]. An additional effect of C-3 substituted oxadienes is the stabilisation of a cisoid conformation. On the other hand, besides the low energy difference between the $LUMO_{oxabutadiene}$ and $HOMO_{dienophile}$ the favourable coefficients at the reaction centers are equally important which can be seen from the Klopman-Salem equation [116–118].

These thoughts do not only count for the 1-oxa-1,3-butadiene, but also for the dienophile. Thus, in an intermolecular cycloaddition with a benzylidenepyrazolone, ethyl vinyl ether reacts about 50 times faster than (*Z*)-1,2-dimethoxyethene and 1,1-diethoxyethene about 2000 times faster than 1,1,2,2-tetramethoxyethene, 3000 times faster than (*E*)-1,2-diethoxyethene, and 5000 times faster than (*Z*)-diethoxyethene [119].

The Diels-Alder reactions with oxabutadienes usually show a high regioselectivity, but in the presence of Lewis acids the regioselectivity is even more enhanced and in addition an increase of the reaction rate is normally observed. Also the stereoselectivity is often improved. Thus, in the intermolecular mode under thermal activation the *endo*/*exo*-selectivity is not very high; if the dienophile is not too bulky as e.g. methyl or ethyl vinyl ether the *endo*-adduct is the major product. However, with *tert*-butyl vinyl ether the *exo*-adduct is formed preferentially. Synthetically highly important is the fact that the *endo*/*exo*-selectivity can be controlled to a high extent by the choice of the Lewis acid passing predominantly either through an *endo*- or an *exo*-transition structure. This again is presumably due to steric reasons. The application of high pressure increases the rate of the cycloadditions and allows the improvement of the *endo*-selectivity in some cases. Of general interest are the transformations which permit the in situ formation of the oxabutadiene in a domino type reaction [3,4] giving access to complex molecules starting from simple compounds in a highly efficient way. As usual, the reaction can be performed in an inter- and intramolecular mode; the latter often shows the higher selectivity.

The thermal reactions of 1-oxa-1,3-butadienes such as acroleine **2-78** with alkenes such as **2-79** usually need relatively harsh conditions (150 °C–250 °C) [120]. As a side reaction polymerisation of the α,β-unsaturated carbonyl compound can take place; addition of radical inhibitors such as hydroquinone or 2,6-di-*tert*-butyl-4-methylphenol can be helpful in avoiding this unwanted transformation. In the described hetero Diels-Alder reaction the cycloadduct **2-80** was obtained which was then transformed into racemic-β-santalene **2-81** (Fig. 2-22).

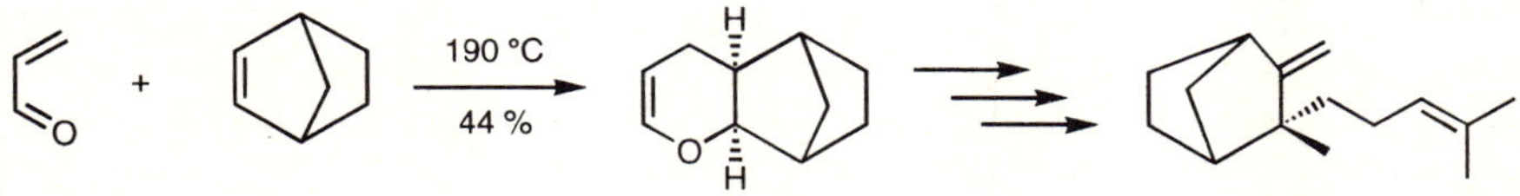

190 °C
44 %

2-78 **2-79** **2-80** **2-81**

β - Santalene

Fig. 2-22

The use of enol ethers as dienophiles improves the reaction, however, still high temperature is needed and *endo*/*exo*-selectivity is low. Thus, cycloaddition of ethyl vinyl ether **2-83** to cyclopentenecarbaldehyde **2-82** gave the cycloadduct **2-84** as a 1:1 mixture which was used for the synthesis of iridoids (Fig. 2-23) [121].

A major breakthrough in the Diels-Alder reaction of oxabutadienes has been accomplished through the introduction of an electron-withdrawing group in the 3-position. In several papers we have demonstrated the usefulness of this concept which found broad acceptance after our discoveries.

Such oxadienes like **2-85** cycloadd to enol ethers like **2-86** already at room temperature with complete regiocontrol and retention of the configuration of the dienophile as well as in many cases with good *endo*-selectivity [122–124]. In the reaction of **2-85** via an *endo-Z-anti* transition structure the major product was the 2,4-*trans*-compound **2-87** (**2-87**:**2-88**=7:1). The (*Z*)-configuration of the 1-oxa-1,3-butadiene in the transition structure is preserved due to the strong hydrogen bond between the carbonyl and the hydroxyl group. Using the corresponding *O*-acetylated compound **2-89a** the *endo*/*exo*-selectivity is reversed to give predominantly the *cis*-1,4-substituted dihydropyran, since now the reaction takes place via an *endo-E-syn*-transition structure. In addition, a dramatic increase in the reaction rate compared to the parent compound occurs.

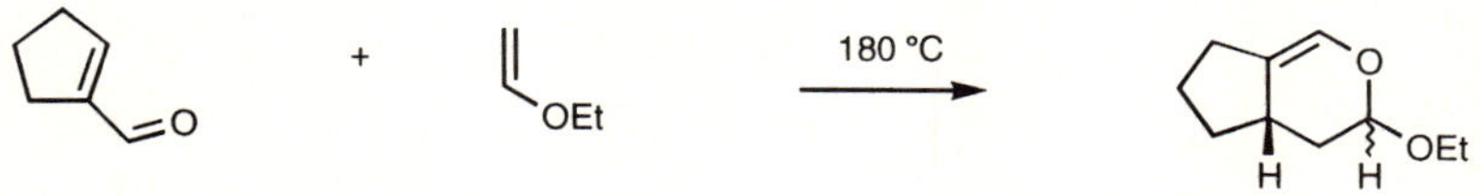

O OEt 180 °C O OEt H H

2-82 **2-83** **2-84**

1 : 1

Fig. 2-23

With the camphanic acid ester derivative **2-89b** a reasonable asymmetric induction was obtained to afford the corresponding dihydropyran after crystallisation in 42% yield as a single diastereomer in enantiopure form (Fig. 2-24) [124].

Interestingly, tricarbonyl compounds such as **2-85,** but not **2-89,** can also undergo a cycloaddition under irradiation to give a different type of dihydropyran.

In the intramolecular mode using either alkylidene- or benzylidene-1,3-dicarbonyl compounds even a simple alkene moiety can act as a dienophile. Depending on the substitution at the dienophile either annulated or bridged cycloadducts can be obtained [3, 4]. The oxadienes e.g. **2-92** are prepared in situ by a Knoevenagel condensation of aldehydes such as **2-90** bearing the dienophile moiety and a 1,3-dicarbonyl compound such as **2-91**; thus, these transformations proceed as domino Knoevenagel hetero Diels-Alder reactions. The method has a broad scope since a multitude of different aldehydes and 1,3-dicarbonyl compounds can be used. Hetero Diels-Alder reactions of oxabutadienes obtained from aromatic aldehydes such as **2-92** lead exclusively to the *cis*-fused cycloadducts like **2-93** (Fig. 2-25) [125], whereas oxabutadienes from aliphatic aldehydes give the *trans*-fused cycloadducts predominantly (~ 98:2) [126]. Applying this protocol a vast array of novel annulated heterocycles can be synthesized in a highly efficient and selective way e.g. reaction of **2-94** and **2-95** yielded exclusively the *cis*-fused tetracycle **2-96** (Fig. 2-26) [127].

On the other hand, oxabutadienes such as **2-97** obtained from aldehydes with a dienophile moiety being unsubstituted at the terminus give bridged compounds **2-98** due to the change of the coefficients at the dienophile moiety (Fig. 2-26a) [128] (see also Sect. 7.1 [490, 492]).

In addition, also spiro compounds can be synthesized using alkylidene-cycloalkenes as a dienophile moiety [56]. Finally, as an immense enlargement of the scope of this protocol, the domino Knoevenagel hetero Diels-Alder reaction can be run as a three component transformation mixing an aldehyde such as **2-99**, a

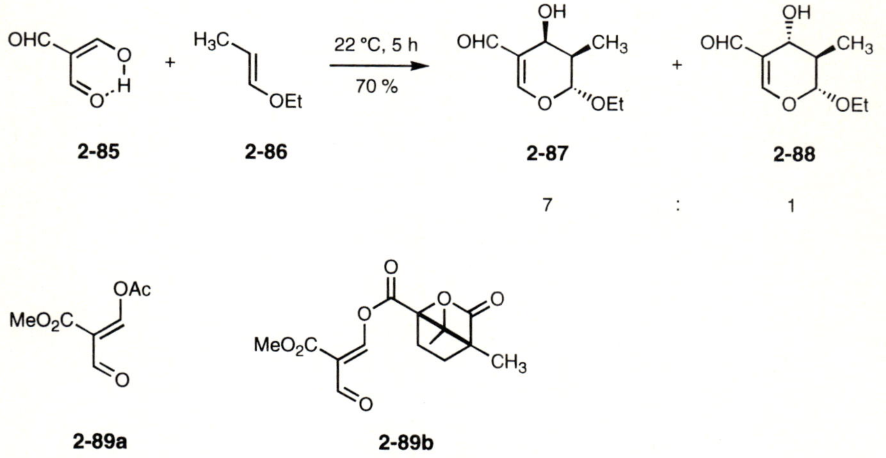

Fig. 2-24

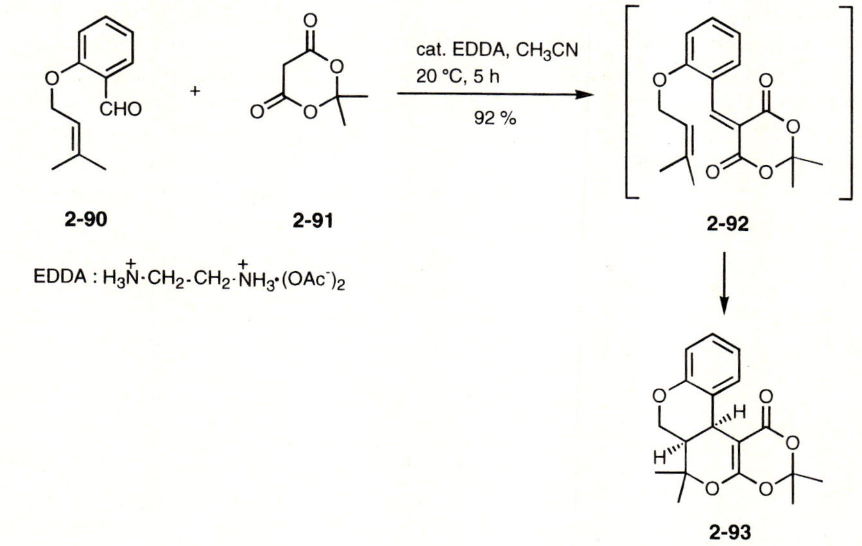

cat. EDDA, CH_3CN
20 °C, 5 h
92 %

2-90 2-91 2-92

EDDA : $H_3\overset{+}{N}\text{-}CH_2\text{-}CH_2\text{-}\overset{+}{N}H_3\cdot(OAc^-)_2$

2-93

Fig. 2-25

cis : *trans* = > 99 : 1

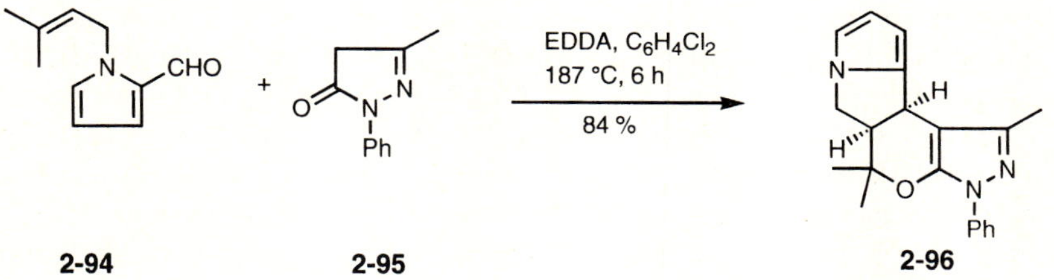

EDDA, $C_6H_4Cl_2$
187 °C, 6 h
84 %

2-94 2-95 2-96

Fig. 2-26

cis : *trans* = > 99 : 1

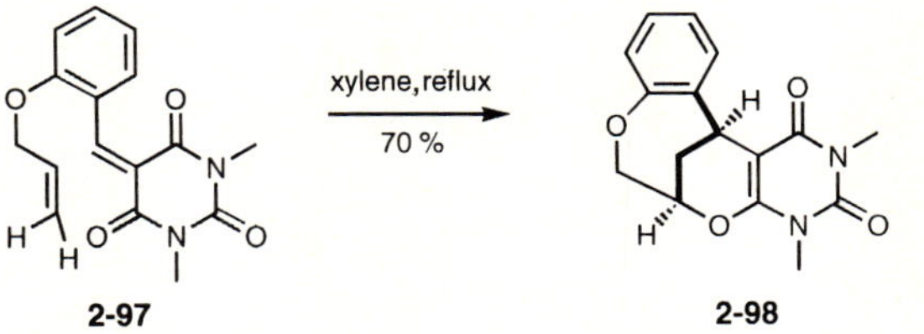

xylene, reflux
70 %

2-97 2-98

+ ortho adduct (1 : 4.5)

Fig. 2-26a

1,3-dicarbonyl compound such as **2-100** and an enol ether such as **2-101** to give the dihydropyran **2-102** (Fig. 2-27) [129].

Unfortunately, formylacetate cannot be applied as 1,3-dicarbonyl compound due to its instability; however, recently we have shown that 4,4,4-trichloro-3-oxobutanal may be used as a formylacetate equivalent, since after the cycloddition the obtained trichloromethylcarbonyl group can easily be transformed into an alkoxycarbonyl group by a base-catalyzed solvolysis with an alcohol [130a]. This concept has been used for the synthesis of secologanin (Sect. 7) [130b].

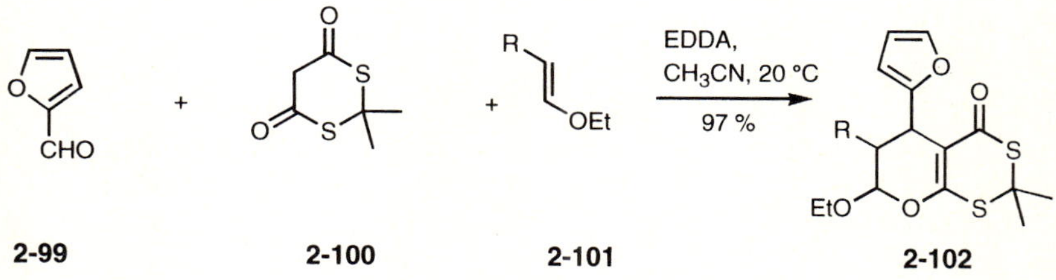

Fig. 2-27

With the highly reactive alkylidene-Meldrum's acid also silyl enol ethers undergo cycloaddition although they are usually less reactive than normal enol ethers [131]. With the even more reactive alkylidene-1,3-diketones such as methylidene-1,3-cyclohexadione **2-103** being obtained in situ by condensation of 1,3-cyclohexadione and formaldehyde a hetero Diels-Alder reaction also with simple alkenes such as **2-104** can be performed in acetic acid [132]. In addition to the cycloadduct **2-105** the corresponding ene product was found (Fig. 2-28). However, usually the chemo- and regioselectivity is high whereas the yields in most cases were only moderate.

In the course of these investigations Hoffmann and his group have also developed novel entries to *t*-butyl 2-methylene-3-oxoalkanoates **2-109a** and 2-methylene-3-oxo-sulfones **2-109b** by oxidation of **2-108a** and **2-108b**, respectively obtained by reaction of the aldehydes 2-106 and acrylate **2-107a** or phenyl vinyl sulfone **2-107b**. The cycloadditions of these oxabutadienes to enol ethers and alkenes proceeded in the expected way (Fig. 2-29) [133].

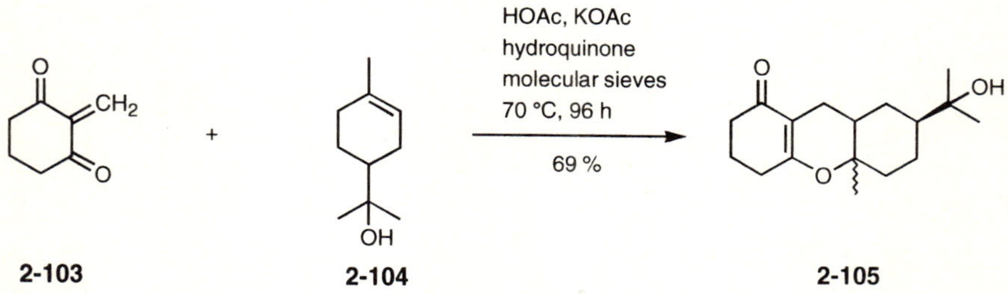

Fig. 2-28

A well established procedure for the synthesis of methylenemalonate **2-111** is the thermolysis of the anthracene derivative **2-110** [134a]. Recently, also a new method for the preparation of the useful methylidene-Meldrum's acid has been described [134b].

For activation of an oxabutadiene also a cyano group at the 3-position can serve as shown by Bogdanowicz-Szwed. Thus, the cycloaddition of 3-cyano-enaminoketones **2-112** with different enol ethers **2-113** in toluene at 100–120 °C for 48–72 h gave the corresponding dihydropyrans **2-114** in good yield (57–90 %), however with low diastereoselectivity (3:1–5:1) with the *cis*-compound always being the major product (Fig. 2-30) [135a]. Earlier we had already shown that enaminoketones with an ester group at C-3 can be used in the cycloaddition with enol ethers to allow an efficient entry to branched amino sugars [135b].

Highly electron deficient *β*,*β*-bis(trifluoroacetyl)vinyl ethers **2-115**, easily prepared by a diacylation of vinyl ethers with trifluoroacetic anhydride, react with electron-rich alkenes in a hetero Diels-Alder reaction smoothly at 20 °C with excellent yield to give the dihydropyran **2-116** (Fig. 2-31). The cycloaddi-

R^1CHO + (alkene)-R^2 → DABCO → **2-108** → Jones oxidation → **2-109**

2-106 **2-107** **2-108** **2-109**

a: $R^2 = CO_2tBu$ b: $R^2 = SO_2Ph$

2-110 → FVT (200 - 250 °C) → **2-211**

Fig. 2-29

2-112 + **2-113** → toluene, 100 - 120 °C, 57 - 90 % → **2-114** (major product)

Fig. 2-30

2-115 + (vinyl ether, OR^2) → 20 °C, 15 min - 24 h, 85 - 98 % → **2-116**

R^1 = Et, *i*Bu
R^2 = Et, Ph

Fig. 2-31

tion shows a high *endo*-selectivity especially when using a vinyl sulfide as dienophile [136].

Reactions of unsymmetrical methylene 1,3-dicarbonyl compounds with enol ethers have been investigated by Yamauchi et al. [137]. As we have mentioned earlier, the α,β-unsaturated ketone moiety in alkylidene-β-ketoesters reacts exclusively as the oxabutadiene. However, high regioselectivity is also observed with mixed alkyl-phenyl-1,3-diketones with exclusive reaction of the aliphatic carbonyl group, whereas in alkylidene-1,3-dicarbonyl compounds bearing an aldehyde and a keto-moiety, the α,β-unsaturated aldehyde reacts preferentially as oxabutadiene, but not exclusively [130a].

Also ynamines can be used in the hetero Diels-Alder reaction of 1-oxa-1,3-butadiene. Novel examples are described by Dell et al. [138] using e.g. the 2-benzylidene-indane-1,3-dione **2-117** and **2-118** to give **2-119** (Fig. 2-32). However, the yields are only modest.

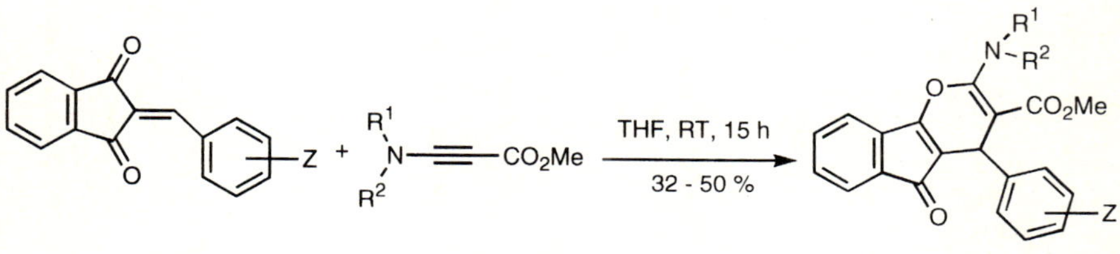

2-117

Z : 4 -NO_2
3 -NO_2
3 -CF_3

2-118

THF, RT, 15 h

32 - 50 %

2-119

Fig. 2-32

Interestingly, the reactivity of oxabutadienes can also be increased by introduction of a thiophenyl group at the 3-position which was first described by Takaki et al. [139] and later employed by Schmidt et al. [140] and us [141]. The use of such oxabutadienes allows an efficient access to 3-aminosugars.

Thus, reaction of **2-120** and **2-121a** gave a 3.9 : 1 mixture of **2-122a** and **2-123a**. **2-122a** could easily be desulfurized to give either the tetrahydropyran **2-124** or the dihydropyran **2-125**. Of importance was the finding that the reaction with the thioenol ether **2-121b** shows an excellent *endo*-selectivity (22 : 1) (Fig. 2-33).

Interestingly, oxabutadienes with an *S*-alkyl group at C-3 such as **2-126**, **2-127** and **2-128** do not react, whereas **2-129** with an *S*-phenyl group again undergoes a cycloaddition. This clearly shows that activation of the oxabutadiene is caused by a kind of conjugation with the phenyl group over the sulfur atom [142]. However, one has to keep in mind that the phenylthio group is only a weak activating moiety compared to electron-withdrawing groups such as CN and CO_2R (Fig. 2-34).

On the other hand, in a series of papers we have shown that 1-oxa-1,3-butadienes **2-130** bearing an electron-withdrawing group such as CN, CO_2R, CCl_3, $CClF_2$ and CF_3 at the 2-position also express a good reactivity, which however, is

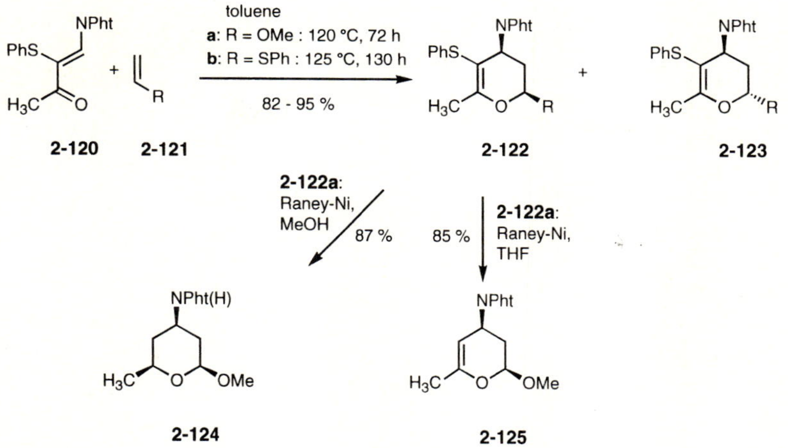

Fig. 2-33

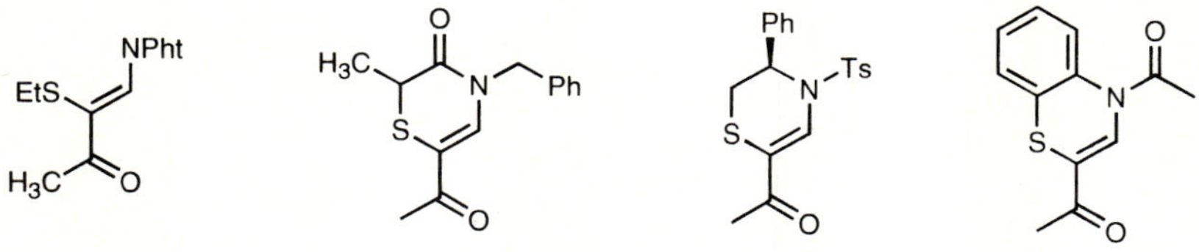

Fig. 2-34

less pronounced than that of oxabutadienes with an electron-withdrawing group at the 3-position. Cycloaddition of **2-130** to electron-rich dienophiles **2-131** gave the dihydropyrans **2-132** and **2-133** in high yield but modest to low selectivity (Fig. 2-35). The highest *endo*-selectivity was obtained employing **2-130** with $R^1 = CO_2$ Me and the lowest with $R^1 = CCl_3$ (see also Sect. 8).

Systematic investigations of the cycloaddition of the 1-oxa-1,3-butadiene **2-134** bearing an ester moiety at the 2-position to enol ethers **2-135** in the presence of Lewis acids were performed by Boger et al. [144] to give the dihydropyrans **2-136** in modest to good yield (Fig. 2-36).

Sera et al. [145] performed similar studies, however using simple alkenes in the presence of $SnCl_4$. The corresponding dihydropyrans **2-139** were obtained generally with high stereoselectivity and partly excellent but also low yield depending on the alkene employed (15–93%). The major product was always the 1,3-*trans* disubstituted compound which is presumably formed via an *exo-E-anti*-transition structure (Fig. 2-37).

Wyler et al. [146a-c] have focused on the hetero Diels-Alder reaction of α,β-unsaturated-acyl cyanides such as **2-140** with ethyl vinyl ether, *N*-methylated uracil and 1-bromo-2-ethoxyethenes **2-141**. In the latter case the dihydropyran

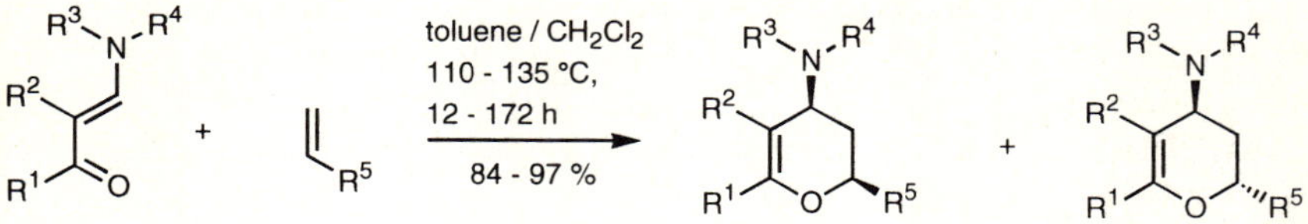

toluene / CH_2Cl_2
110 - 135 °C,
12 - 172 h
84 - 97 %

2-130 2-131 2-132 2-133

ds =1.5 : 1 → 6.6 : 1

R^1 : CO_2Me, CCl_3, $CClF_2$, CF_3
R^2 : H, Br
R^3 : PhCO; R^4 : H; R^3R^4: Pht
R^5 : OEt, SPh

Fig. 2-35

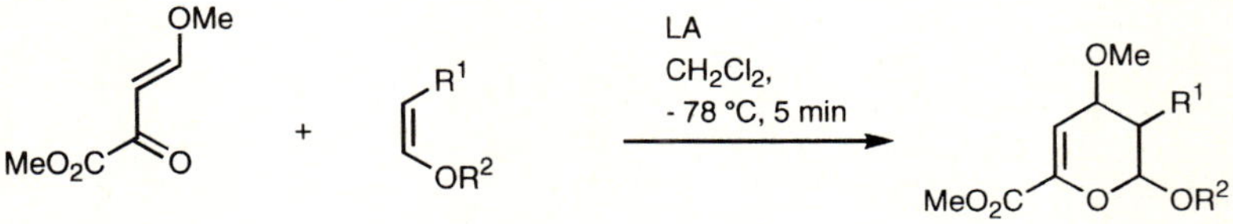

LA
CH_2Cl_2,
- 78 °C, 5 min

2-134 2-135 2-136

a: R^1 = H, R^2 = Et
b: R^1 = CH_3, R^2 = Bn

2-135	LA	**2-136** *endo / exo*	yield [%]
a	$EtAlCl_2$	0.8 : 1	75
a	$TiCl_4$	1 : 3.4	61
b	$EtAlCl_2$	6 : 1	46

Fig. 2-36

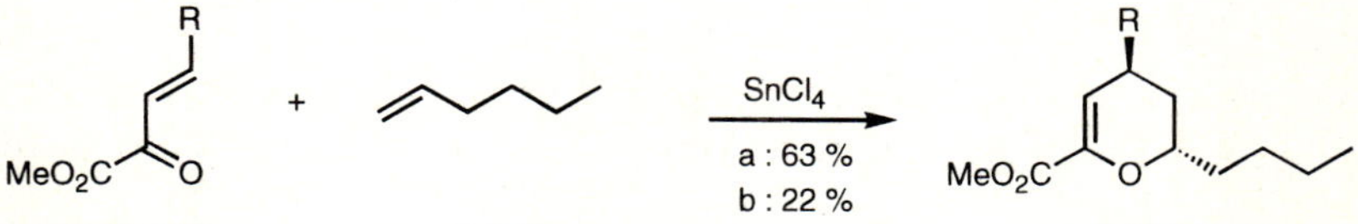

$SnCl_4$
a : 63 %
b : 22 %

2-137 2-138 2-139

a : R = Ph
b : R = CH_3

Fig. 2-37

2-142 was obtained, which could be transformed into stable pyrans by elimination of hydrogen bromide. The yields were good, however, the selectivity, except for the reaction of **2-140c** and **2-141a**, was rather low (Fig. 2-38).

In a similar way, also other groups have investigated the cycloaddition of α,β-unsaturated acyl cyanides [146d].

Acyl ketenes also react with a variety of dienophiles such as enol ethers to give the corresponding 2-alkoxy-2,3-dihydro-4-pyranones [147].

A versatile approach to spiro-oxacycles is the use of cyclic α-methylene enol ethers employed by us in an efficient and short enantioselective total synthesis of the mycotoxin talaromycin B (see Sect. 7.1). Later Pale and Vogel [148] employed the same protocol for the preparation of spiroacetals **2-145** using e.g. acrolein **2-78**, methyl vinyl ketone and 2-pentenal, respectively with the enol ether **2-143** (Fig. 2-39). In most cases the yields were only modest, however, reaction of **2-143** and **2-78** in benzene in the presence of the mild Lewis acid $ZnCl_2$ gave **2-145** in 70% yield as a single adduct.

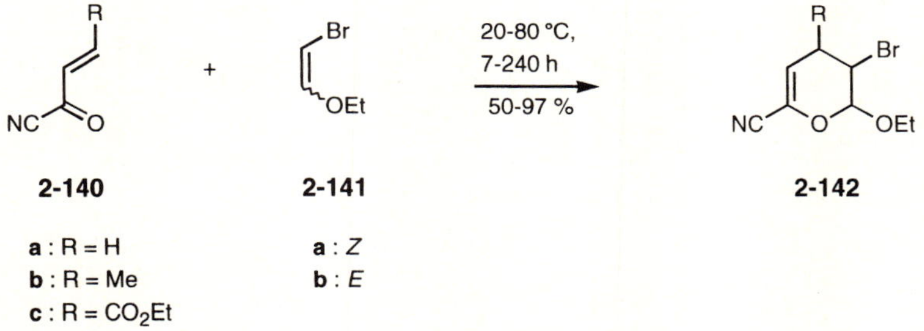

20-80 °C, 7-240 h, 50-97 %

2-140 **2-141** **2-142**

a : R = H
b : R = Me
c : R = CO_2Et

a : *Z*
b : *E*

Fig. 2-38

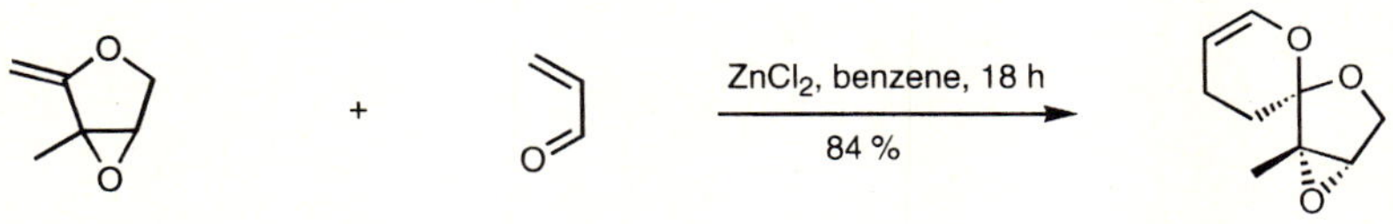

$ZnCl_2$, benzene, 18 h, 84 %

2-143 **2-78** **2-145**

Fig. 2-39

Novel complex heterocycles such as **2-148** can easily be obtained using the hetero Diels-Alder reaction of enol ethers like **2-147** and **2-146** as 1-oxa-1,3-butadiene (Fig. 2-40) [149].

For the reaction of non-activated 1-oxa-1,3-butadienes several different Lewis acids have been developed. Thus, the $MoO_2(acac)_2$-catalyzed reaction of methacrolein with enol ethers gave dihydropyrans in 58% yield at 100 °C [150].

Especially successful is the use of a Lewis acid in those cases, where a second chelating group exists in the molecule. Wada et al. [151] have shown that (*E*)-2-oxo-1-sulfonyl-3-alkenes **2-149** cycloadd in hetero Diels-Alder reactions to enol

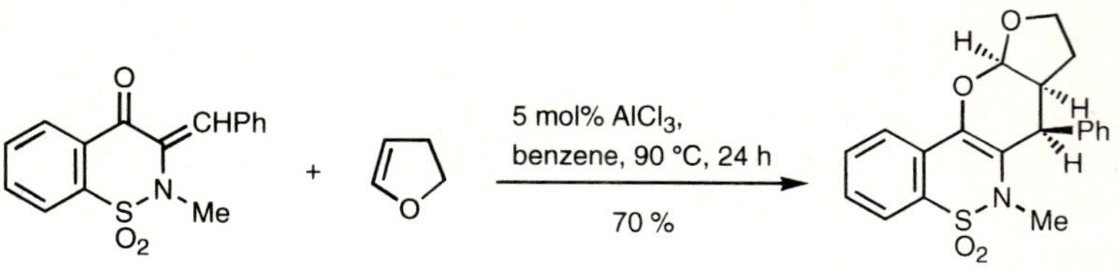

5 mol% $AlCl_3$, benzene, 90 °C, 24 h

70 %

2-146 2-147 2-148

major product

other enol ethers used: ethyl vinyl ether, dihydropyran

endo : *exo* = 85 : 15

Fig. 2-40

ethers **2-150** in the presence of Lewis acids like $Eu(fod)_3$ or $TiCl_2(i\text{-}PrO)_2$ with excellent selectivity to give the *endo*-products **2-151**. A further advantage of this approach is the possibility to manipulate the obtained dihydropyrans **2-151** at the side chain via formation of a sulfonyl-stabilized carbanion (Fig. 2-41).

Alkylidene-3-indolones **2-152** react with ethyl vinyl ether **2-83** to give the corresponding dihydropyrans **2-153** in the presence of $Yb(fod)_3$ (Fig. 2-42) [152].

Also 1-ethoxy-1,2-propadiene (ethoxyallene) can be used as a dienophile in hetero Diels-Alder reactions. In this case the reaction was performed under dry state adsorption conditions on silica gel [153].

Desimoni and Righetti have been thoroughly investigating the effect of solvents, acid catalysis and salts on hetero Diels-Alder and ene reactions of 1-oxa-1,3-butadienes for a long time [154–156]. Recently, for the cycloaddition of **2-154** and ethyl vinyl ether **2-83** in the presence of lithium perchlorate in different solvents as diethyl ether, acetonitrile, acetone, methanol, *iso*-propanol to give

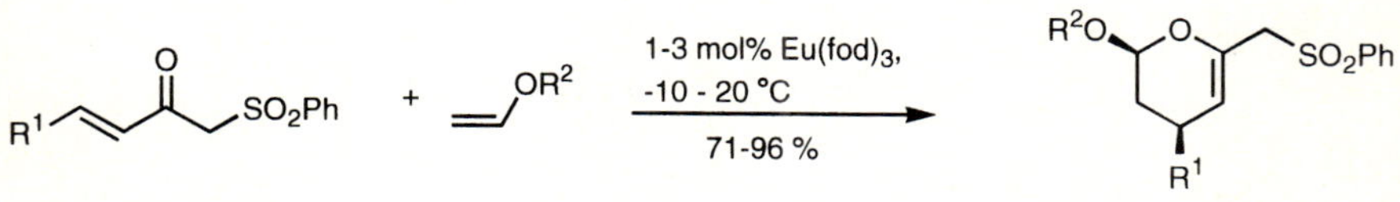

1-3 mol% $Eu(fod)_3$, -10 - 20 °C

71-96 %

2-149 2-150 2-151

R^1 = Me, *i*Pr, Ph

R^2 = Et, *i*Bu, Ph

>99 : 1

Fig. 2-41

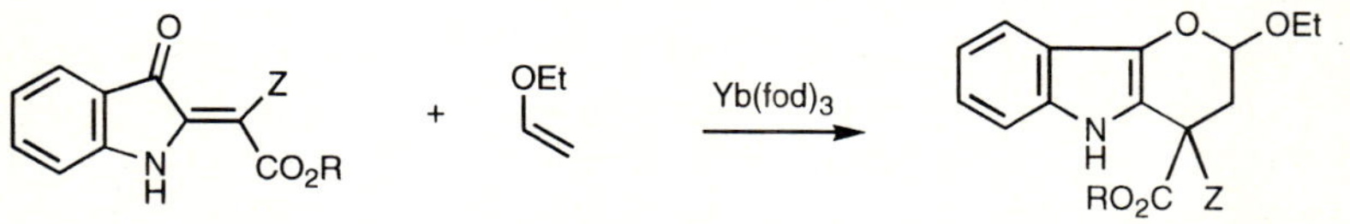

$Yb(fod)_3$

2-152 2-83 2-153

Fig. 2-42

2-155 a linear increase of the relative rate by increasing the molar fraction of the salt was observed with a k_{rel} value of 850 ± 80 (Fig. 2-43). There is a slight increase of the *endo*-selectivity in diethyl ether (without $LiClO_4$: 88/12, 1M $LiClO_4$: 95/5) but no effect in methanol. The results can be rationalized by the lithium cation acting as a Lewis acid [80] and, as already mentioned, not as the effect of an "internal pressure" [157].

A major problem in the reaction of α,β-unsaturated carbonyl compounds and alkenes proves to be the competition between hetero Diels-Alder and ene reactions. Intramolecular cycloadditions of 1,6- and 1,7-dienes with ester and cyano groups at the double bond yield the ene product nearly exclusively, but with alkylidene- and benzylidene-ketoesters and 1,3-diketones the Diels-Alder reaction is preferred under thermal conditions, however under Lewis acid catalysis also ene reactions occur [12].

In a series of papers Desimoni and Righetti [158-160] have now shown that the addition of salts such as lithium perchlorate and magnesium perchlorate not only accelerates the reaction, but also has a high influence on the ratio of ene and Diels-Alder products. Reaction of the benzylidene-1,3-diketone **2-156** with $LiClO_4$ at 25 °C for 3 days a 85:15 ratio of **2-157** and **2-158** in 70% yield and with $Mg(ClO_4)_2$ at 25 °C for 20 h a 7:93 ratio in 100% yield was obtained (Fig. 2-44). However, as expected the corresponding benzylidene-malonate gives the ene product exclusively under all conditions.

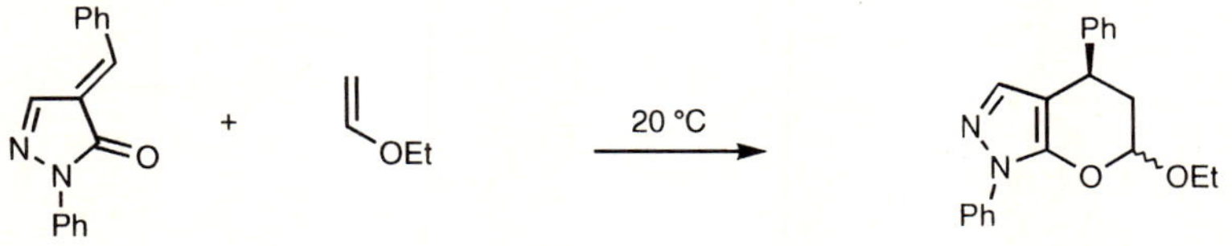

2-154 + **2-83** —(20 °C)→ **2-155**

Fig. 2-43

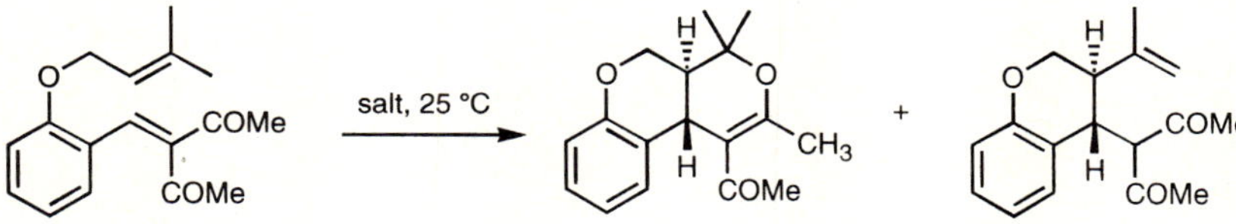

2-156 —(salt, 25 °C)→ **2-157** + **2-158**

	2-157		2-158
$LiClO_4$:	85	:	15
trans : *cis* =	2 :1		>95 : 5
$MgClO_4$:	7	:	93
trans : *cis* =	~1 :1		> 99 : 1

Fig. 2-44

Recently a remarkable chemo- and enantioselectivity has been found using the *trans*-(4,5-diphenyloxazoline)-magnesium perchlorate complex **2-161** for the transformation of **2-159** to yield the ene product **2-160** with 88% *ee* in a 89:11 preference (Fig. 2-45).

In recent years several highly efficient asymmetric cycloadditions employing chiral oxabutadienes have been developed. The use of chiral dienophiles was less satisfying as shown for the cycloaddition of **2-162** to **2-120** to give the dihydropyran **2-163** (Fig. 2-46) [161c].

However high *endo*-selectivity with excellent asymmetric induction was obtained by us with oxabutadienes **2-164** bearing an oxazolidinone moiety derived from *tert*-leucine and enol ethers **2-165.** The results are surprising, if one considers that the inducing stereogenic center is five atoms away (1,6-induction). In the presence of Me_2AlCl **2-166** was obtained nearly exclusively; whereas with TMSOTf the *endo*-cycloadduct **2-167** was the major product showing the

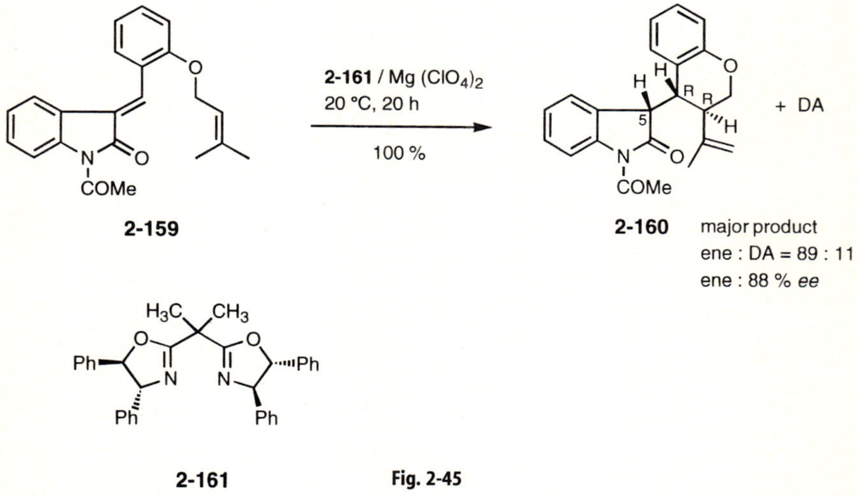

Fig. 2-45

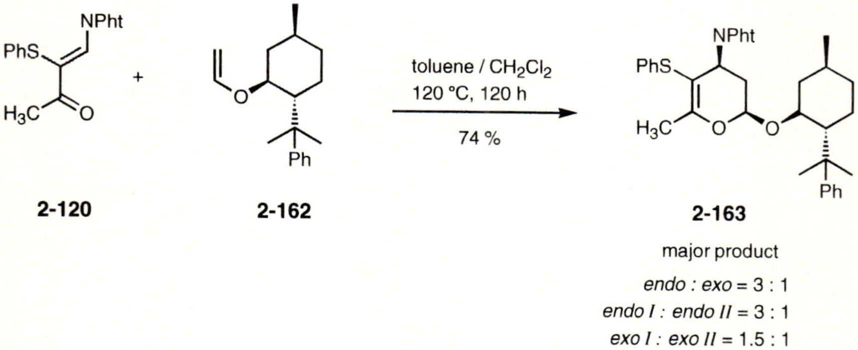

Fig. 2-46

opposite absolute configuration of the dihydropyran moiety [161]. Thus, a reversal of the facial selectivity is possible by just changing the Lewis acid to give access to both enantiomers of the cycloadduct with the same auxiliary. The results could be explained from calculations [162]. Using bidental Lewis acids such as Me_2AlCl a chelate **2-171** is formed with the *tert*-butyl group facing down. However, with monodental Lewis acids such as TMSOTf, the oxenium ion **2-170** with the opposite orientation of the *tert*-butyl group is presumably the intermediate. Also, a reversal of the *endo*/*exo*-selectivity is possible using $SnCl_4$ as Lewis acid to give **2-168** as the major product again with high induced diastereoselectivity for the *endo*- and *exo*-adduct (Fig. 2-47) [163].

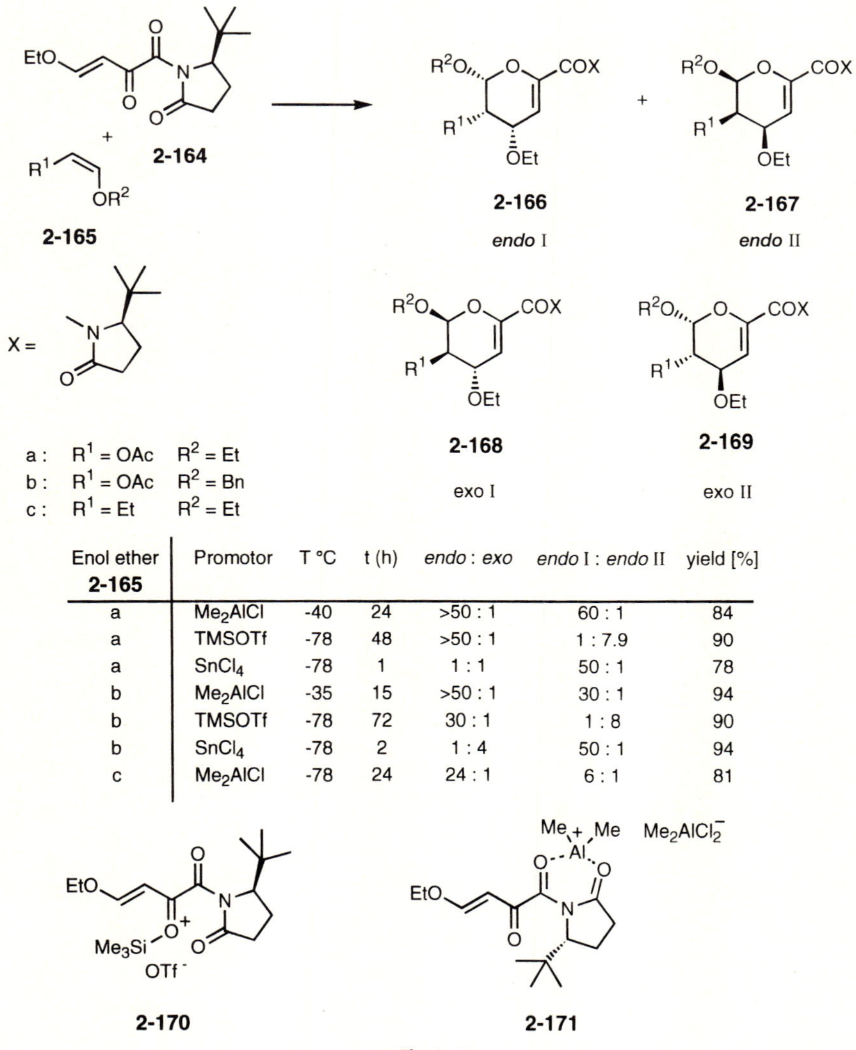

Enol ether 2-165	Promotor	T °C	t (h)	*endo* : *exo*	*endo* I : *endo* II	yield [%]
a	Me_2AlCl	-40	24	>50 : 1	60 : 1	84
a	TMSOTf	-78	48	>50 : 1	1 : 7.9	90
a	$SnCl_4$	-78	1	1 : 1	50 : 1	78
b	Me_2AlCl	-35	15	>50 : 1	30 : 1	94
b	TMSOTf	-78	72	30 : 1	1 : 8	90
b	$SnCl_4$	-78	2	1 : 4	50 : 1	94
c	Me_2AlCl	-78	24	24 : 1	6 : 1	81

Fig. 2-47

An excellent asymmetric induction has also been observed by Snider et al. [164] in a hetero Diels-Alder reaction with the *N*-crotonyl oxazolidinone **2-172** which had already been used by Evans for the all carbon cycloaddition [164b]. Reaction of isobutene with **2-172** in the presence of Me_2AlCl for 40 h at −30 °C in CH_2Cl_2 provided a mixture of the alcohol **2-174** and the lactone **2-175** via the primarily formed cycloadduct **2-173**. Treatment of the mixture with sodium carbonate gave the lactone **2-175** as a pure enantiomer (Fig. 2-48).

Ephedrine derived benzylidene-oxazepandiones **2-178** have also been proven as very effective chiral 1-oxa-1,3-butadienes in asymmetric hetero Diels-Alder reactions. Knoevenagel reaction of the aldehyde **2-176** with the oxazepanedione **2-177** gave the (*Z*)-benzylidene derivative **2-178** which undergoes a hetero Diels-Alder reaction in the presence of Et_2AlCl to afford the cycloadduct **2-179** in good yield and excellent selectivity (78%, *de* >98%) [165]. The auxiliary could be removed by treatment with acid and base to give the enantiopure lactone **2-180** (Fig. 2-49). Interestingly, the cycloaddition takes place in an *exo-Z-syn* fashion with an attack at the oxadiene *syn* to the bulkier groups at the stereogenic centers in **2-178**. The reason for this unusual result is the conformation of the benzylidene-oxazepandione which prohibits the attack from the *Si-face* due to two hydrogens in **2-178** at the lower face [165].

Kaneko and Sato [166–169] have also developed several very useful chiral 1-oxa-1,3-butadienes such as the enantiopure benzylidene-1,3-dihetero-4,6-dioxocyclohexanes **2-181** and **2-182** and the 5,6-dihydropyran-2,4-dione **2-183** (Fig. 2-50) obtained by Knoevenagel condensation of the corresponding 1,3-dicarbonyl compound with benzaldehydes. The hetero Diels-Alder reaction of **2-184** with ketene acetal proceeded with good yields and high selectivity to give

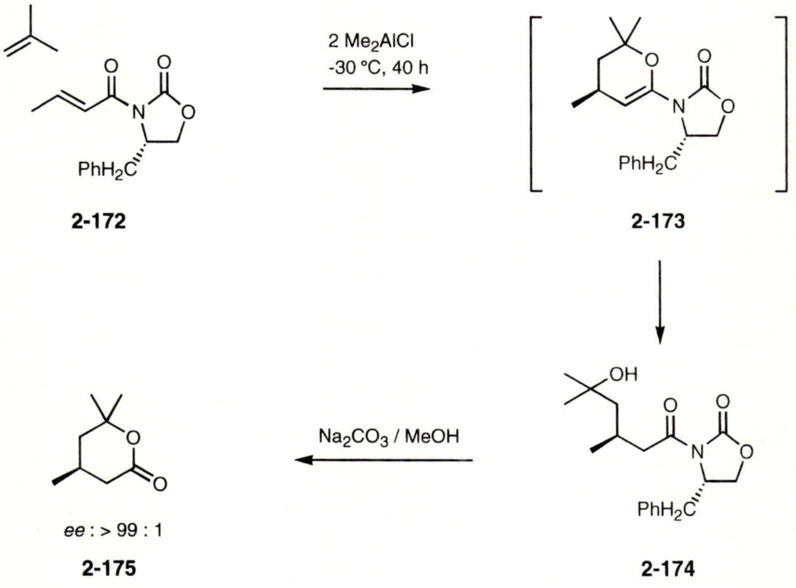

Fig. 2-48

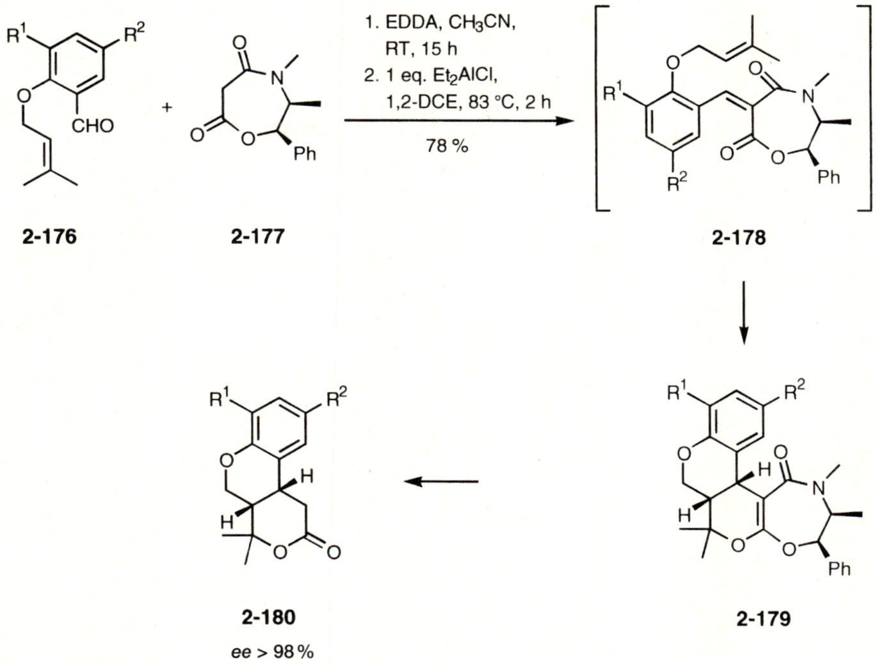

1. EDDA, CH_3CN, RT, 15 h
2. 1 eq. Et_2AlCl, 1,2-DCE, 83 °C, 2 h

78 %

2-176 2-177 2-178

2-180 2-179

ee > 98 %

Fig. 2-49

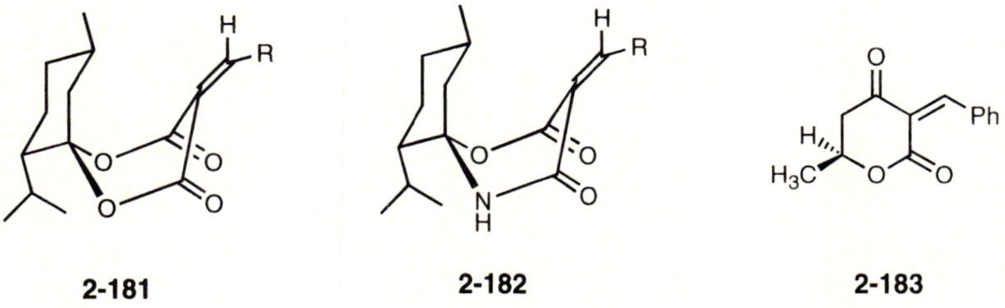

2-181 2-182 2-183

Fig. 2-50

the adduct **2-185** which could easily be transformed in three steps into the 5-ketoester **2-186** with a stereogenic center at C-3 (Fig. 2-51).

A high asymmetric induction in intramolecular hetero Diels-Alder reactions was found using chiral 1-oxa-1,3-butadienes with a stereogenic center in the tether [54]. Such compounds can easily be obtained by a Knoevenagel condensation of a 1,3-dicarbonyl compound such as *N,N*-dimethylbarbituric acid with a chiral aldehyde bearing a dienophile moiety [169a] (Scheme 2-3). With the stereogenic center in α-position relative to the oxadiene or dienophile moiety an excellent induced diastereoselectivity is obtained for the nearly exclusively formed *trans*-cycloadduct (simple diastereoselectivity = 97.9:2.1 and 98.3:1.7,

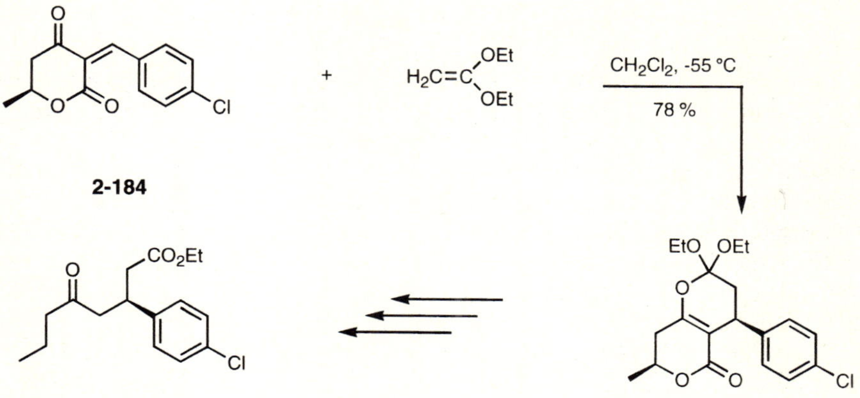

CH_2Cl_2, -55 °C

78 %

2-184

2-186

Fig. 2-51

2-185

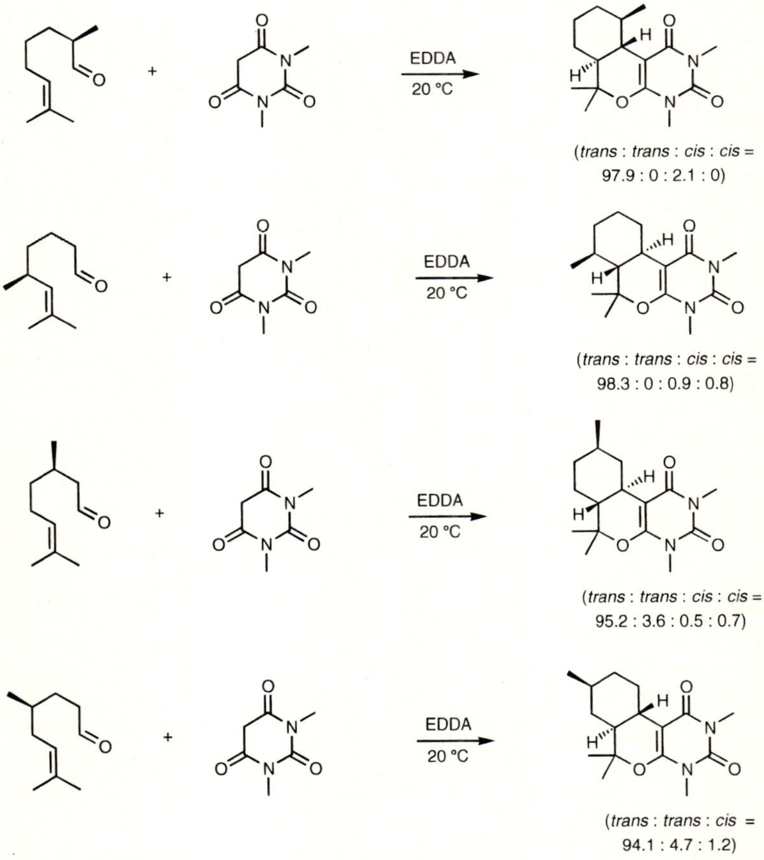

EDDA 20 °C

(*trans* : *trans* : *cis* : *cis* = 97.9 : 0 : 2.1 : 0)

EDDA 20 °C

(*trans* : *trans* : *cis* : *cis* = 98.3 : 0 : 0.9 : 0.8)

EDDA 20 °C

(*trans* : *trans* : *cis* : *cis* = 95.2 : 3.6 : 0.5 : 0.7)

EDDA 20 °C

(*trans* : *trans* : *cis* = 94.1 : 4.7 : 1.2)

Scheme 2-3. Intramolecular hetero Diels-Alder reactions with a stereogenic center in the tether

induced diastereoselectivity >99:1). The observed high stereoselectivity can be explained by the sp^2-geminal effect, which is a type of 1,3-allylic strain [169b].

With the stereogenic center in the β-position the induced diastereoselectivity is controlled by the preferential equatorial orientation of the substituent in a chair-like transition structure. However, the selectivity is higher (96.3:3.7) than could be anticipated from a simple comparison with the ratio of the equatorially and axially orientated methyl group in methylcyclohexane (95:5).

Also, chiral α-sulfinyl-α,β-unsaturated ketones **2-187** have been employed in hetero Diels-Alder reactions, however, the observed diastereoselectivities were less satisfying. Using Et_2AlCl as the best promotor 81.7% *de* and with $ZnCl_2$ 30.5% *de* was found (Fig. 2-52) [170a].

As expected, the major product was the *trans*-annulated compound **2-188** which should be formed via an *exo-E-anti*-transition structure with an attack of the dienophile *anti* to the bulky tolyl group. An *endo-Z-anti*-orientation which would also lead to the *trans*-product can be excluded because of its strain [54]. The reaction of the corresponding α'-sulfinyl-α,β-unsaturated ketone **2-189** displayed a much lower induced diasteroselectivity (Fig. 2-53) [170b].

An interesting approach to directly linked C-disaccharides **2-191** was developed by Dondoni et al. [171] via a hetero Diels-Alder reaction of sugar derived 1-oxa-1,3-butadienes as **2-190** bearing a thiazole moiety at position 2 as activating group (Fig. 2-54).

The cycloaddition with ethyl vinyl ether proceeded with excellent yield either under thermal conditions (80 °C, 5 days) or in the presence of $LiClO_4$ (20 °C, 17 h). In the former case a reasonable *exo*/*endo* and induced diastereoselectivity was observed, however using $LiClO_4$ the asymmetric induction was null.

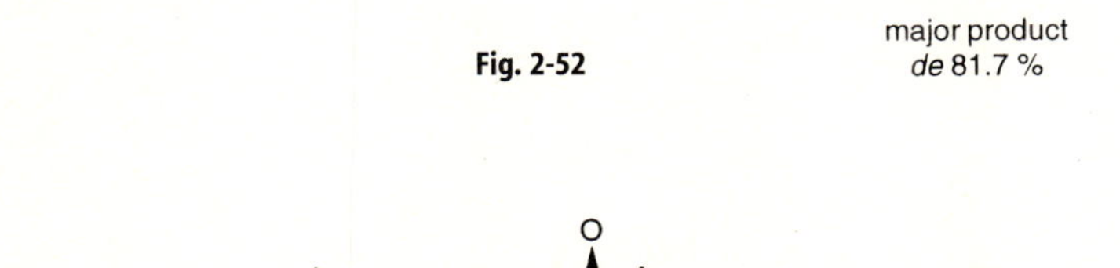

Fig. 2-52

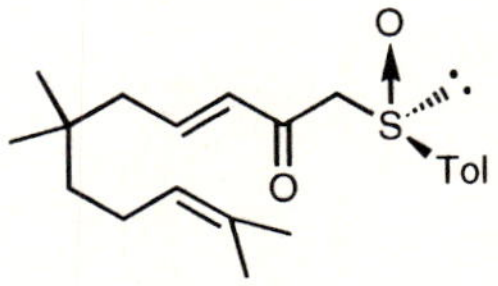

Fig. 2-53

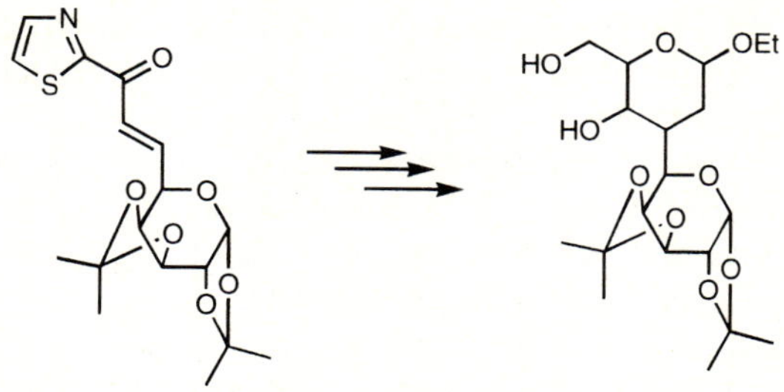

Fig. 2-54

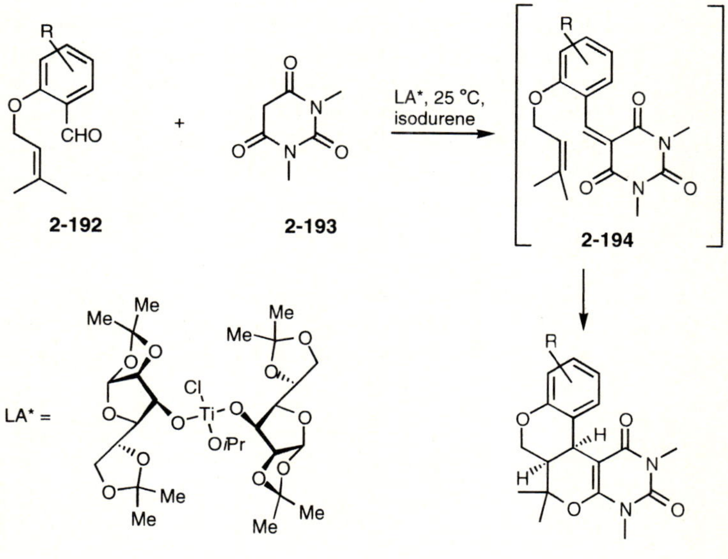

Substrate 2-192	R	Product 2-195	yield [%]	*ee* [%]
a	3-OMe	**a**	68	75
b	4-OMe	**b**	72	80
c	5-OMe	**c**	86	88
d	6-OMe	*ent*-d	61	30

Fig. 2-55

Sugar derived enals like 2- and 3-formyl hex-1- and hex-2-enopyranosides have also been used for the cycloaddition with enol ethers in the presence of $Eu(fod)_3$ as a catalyst with good yields and high selectivity [172].

The enantioselective hetero Diels-Alder reaction of 1-oxa-1,3-butadienes using chiral non-racemic Lewis acids is a widely unexplored field. The first successful example was the intramolecular cycloaddition of the heterodiene **2-194**, obtained in situ from **2-192** and **2-193** by Knoevenagel reaction. In this transformation the novel diacetone glucose derived Lewis acid **2-196** promotes the condensation and the cycloaddition to give the products **2-195** via **2-194** (Fig. 2-55). The reaction is completely *cis*-selective and yields the 5-methoxy substituted tetracycle **2-195c** with 88% *ee* [173]. The reaction is highly solvent dependent with 88% *ee* in isodurene, 86% *ee* in toluene, 72% *ee* in tetrahydrofuran and 34% *ee* in dichloromethane. In addition, it shows a highly interesting temperature curve with 88% *ee* at 25 °C and nearly 0% *ee* at –50 °C and + 100 °C. The temperature dependence is in agreement with the principal of isoinversion [174]. Astoundingly, in the reaction of **2-192d** and **2-193** *ent*-**2-195d** is the major cycloadduct.

Recently, Wada et al. [175, 176] have observed an excellent enantioselectivity for the intermolecular hetero Diels-Alder reactions of (*E*)-2-oxo-1-phenylsulfonyl-3-alkenes **2-197** – **2-199** with enol ethers **2-200** to give the dihydropyrans **2-201** – **2-203** using the Narasaka catalyst **2-204.** The best results were obtained with the *iso*-propyl vinyl ether **2-200c** (Fig. 2-56).

The titanium dibromo catalyst **2-204** gave in all cases better yields and a higher selectivity than the corresponding dichloro complex. The results can be rationalized by the formation of a chelate between the catalyst and the carbonyl and the sulfonyl group.

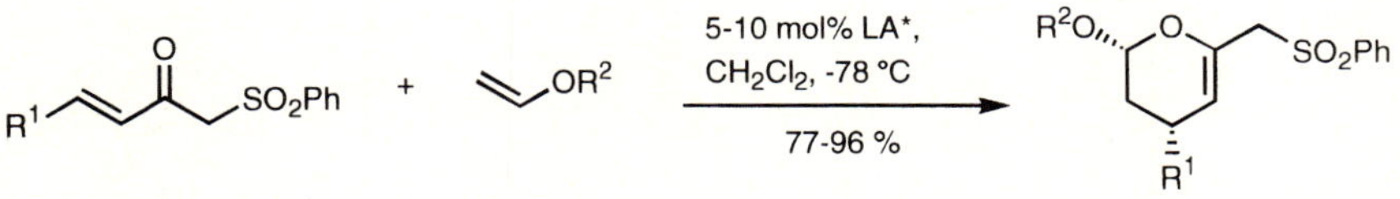

product	yield [%]	% *ee*
2-201a	91	59
2-201b	96	74
2-201c	90	97
2-202c	88	86
2-203c	77	97

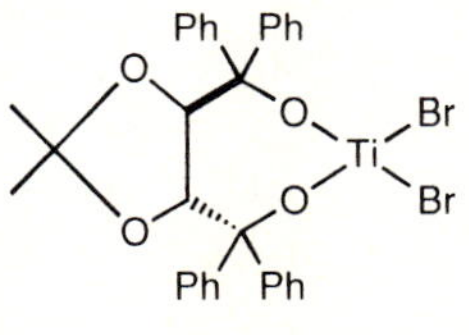

Fig. 2-56

3
Aza Diels-Alder Reactions

3.1
Reactions with C=N Dienophiles

The formation of tetrahydropyridines by reaction of a suitable diene with an imino dienophile is a reaction known since more than half a century [177] and has been intensively studied. In general, the imines react as the electron-deficient component and their reactivity strongly depends on the electron density which may be tuned by activating or deactivating moieties. However, exceptions from this rule are possible as found by Padwa et al. [178]. They described cycloadditions of imines to bis(phenylsulfonyl)-1,3-butadienes.

Aza Diels-Alder reactions of nonactivated imines have been investigated with regard to the effects of electronically neutral substituents and the influence of Lewis acids [179, 180]. Some of the synthetic applications described below nevertheless take use of the higher reactivity of iminium ions or imines activated by electron-withdrawing substituents, respectively.

The more recent work on this area deals predominantly with the asymmetric induction in aza Diels-Alder reactions in order to develop a novel powerful tool for the stereoselective synthesis of biologically active compounds. Thus, Waldmann et al. demonstrated the utility of chiral imines derived from enantiopure amino acids by obtaining the cycloadduct **3-3** in very good diastereoselectivity from imine **3-1** and Brassard's diene **3-2** (Fig. 3-1) [181].

A similar approach to the synthesis of tetracyclic indole alkaloid derivatives has been described [182], and the use of reactive chiral iminium ions allows the realisation of stereoselective aza Diels-Alder reactions even in aqueous solution [183, 184]. Nevertheless it should be noted that reactions of electron-rich dienes with imines e.g. derived from amino acids do not necessarily proceed via a Diels-Alder mechanism. They may as well undergo a domino-Mannich-Michael sequence which also efficiently leads to useful nitrogen heterocycles [185–188].

An elegant approach to indolizidine and quinolizidine derivatives using imines derived from sugars has been presented by Herczegh et al. [189–191]. The imine generated in situ from aldehyde **3-4** derived from D-glucose reacted smoothly with Danishefsky's diene to form **3-5** which was easily transformed to the aldehyde **3-7**. Hydrogenolysis under acidic conditions directly yielded the castanospermine analogue **3-6** (Fig. 3-2).

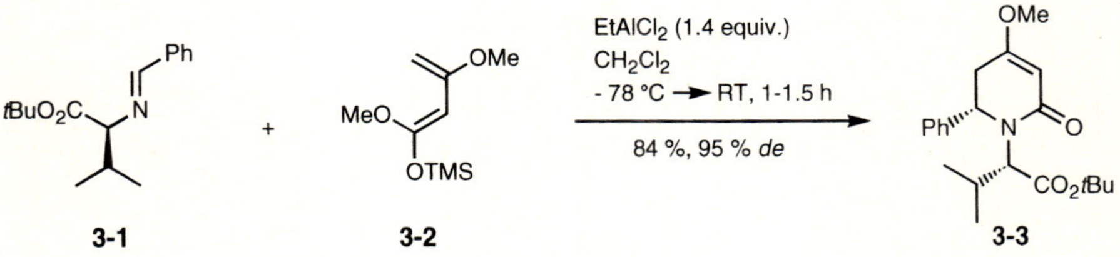

tBuO2C, N, Ph **3-1** + MeO, OMe, OTMS **3-2** —[$EtAlCl_2$ (1.4 equiv.), CH_2Cl_2, - 78 °C → RT, 1-1.5 h; 84 %, 95 % *de*]→ OMe, Ph, N, O, CO_2tBu **3-3**

Fig. 3-1

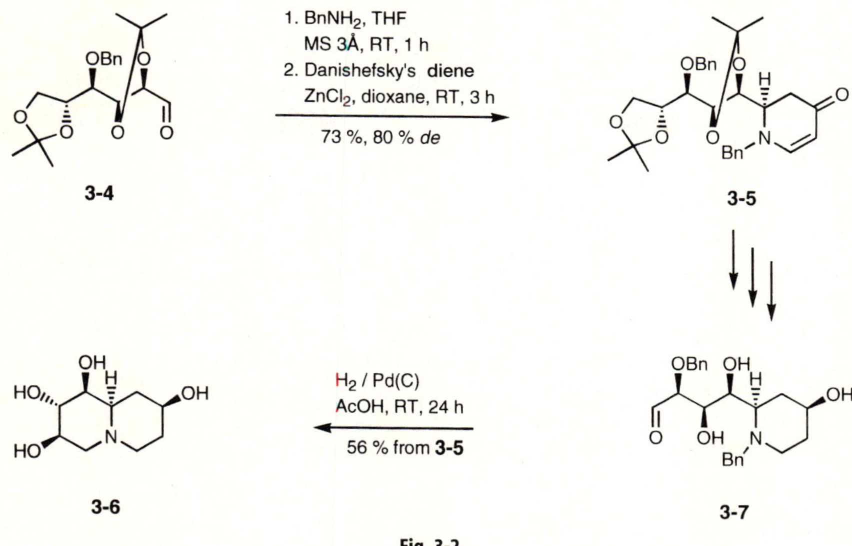

Fig. 3-2

Very recently, chiral tricarbonylchromium complexes have been introduced as novel chiral auxiliaries for aza Diels-Alder reactions [192, 193]. Using the brominated imine **3-8**, Kündig's group was successful in efficiently generating enantiopure polycyclic compounds such as **3-10** by cycloaddition of **3-8** to 1-methoxy-3-trimethylsilyloxy-1,3-butadiene (Danishefsky's diene), subsequent radical cyclisation of the cycloadduct **3-9** and oxidative metal removal from **3-11** (Fig. 3-3).

$SnCl_4$ (1.2 equiv.)
THF, -78 °C → RT, 15 h
63 %, 96 % *de*

3-8 **3-9**

Bu_3SnH
AIBN
benzene reflux.
63 %
>95 % *de*

light, air
CH_3CN
95 %, >98 % *ee*

3-10 **3-11**

Fig. 3-3

Numerous further chiral imines activated by electron-withdrawing substituents have been investigated in order to carry out stereoselective aza Diels-Alder reactions. In these studies, Bailey et al. have recently introduced the use of two inducing stereocenters in the imine. This approach proved to yield excellent diastereoselectivities; thus, imine **3-12** bearing a (*R*)-8-phenylmenthyl auxiliary gave the essentially pure cycloadduct **3-13** upon hetero Diels-Alder reaction with cyclopentadiene (Fig. 3-4) [194–196].

Chiral Lewis acids have been employed by Yamamoto et al. [197–199] in order to carry out enantioselective aza Diels-Alder reactions starting from achiral substrates; however, these transformations required stoichiometric amounts of the chiral mediator **3-16** which was generated in situ from (*R*)-binaphthol and triphenylborate. The best results were obtained with the pyridine derivative **3-14** which afforded the desired cycloadduct **3-15** in high optical purity (Fig. 3-5). Using chiral imines, the authors found a high level of double asymmetric induction, and this methodology could be applied to the enantioselective total synthesis of two piperidine alkaloids.

Further studies concerning aza Diels-Alder reactions of *N*-sulfonyl imines have been carried out by Holmes et al. [200] and Whiting et al. [201, 202], and the utility of glyoxylato imines for the synthesis of cyclic amino acids has been investigated by Stella's group [203, 204]. An extensive study concerning intramole-

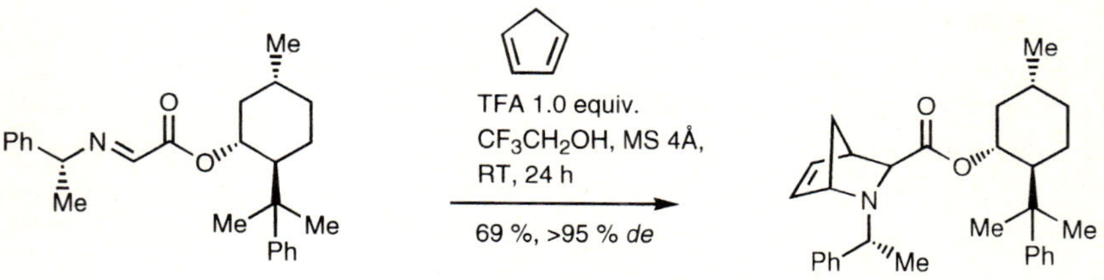

3-12 TFA 1.0 equiv., CF_3CH_2OH, MS 4Å, RT, 24 h; 69 %, >95 % *de* **3-13**

Fig. 3-4

3-14 (*R*)-**3-16** (1 equiv.), CH_2Cl_2, - 78 °C, 5 h; 71 %, 90 % *ee* **3-15**

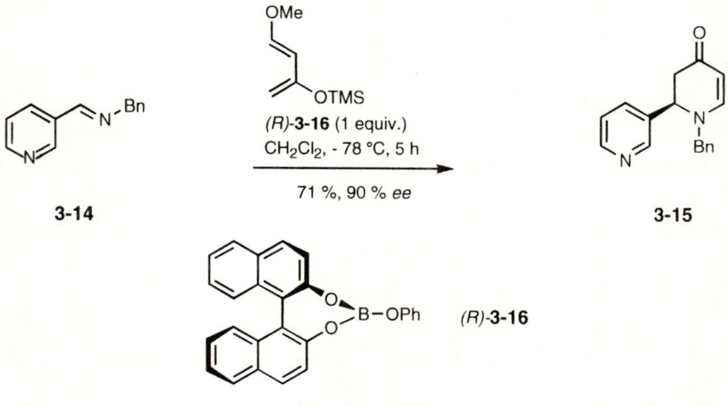

(*R*)-**3-16**

Fig. 3-5

cular cycloadditions of imines activated by acyl and ester moieties has been carried out by Shea et al. [205].

The aza Diels-Alder reactions of chiral α-alkoxy imines described by Midland et al. [206] proceeded only in the presence of strong Lewis acids due to the deactivation of the dienophile by the electron-donating substituent, but with suitable substrates, very high diastereoselectivities could be achieved. It is as well possible to use chiral dienes in such transformations as asymmetric inductors; this kind of asymmetric aza Diels-Alder reactions has recently been investigated by Barluenga and his coworkers [207] by employing chiral 2-amino-1,3-butadienes [208]. A noteworthy application of high pressure in aza Diels-Alder reactions with imino dienophiles is the synthesis of carbocyclic nucleosides from 2-iminomalonates by Katagiri et al. [209]. 1,3-Oxazine derivatives are avilable from imines and heterocumulenes by cycloaddition to dipivaloylketene [210]. The retro-aza Diels-Alder reaction has been studied by Grieco et al. [211, 212] in order to release reactive iminium ions. An application of this technique to peptide chemistry has also been developed [213]. Finally, Katritzky et al. have shown that reactive iminium ions formed from *N*-(α-aminoalkyl)benzotriazoles are suitable imino dienophiles for aza Diels-Alder reactions [214].

3.2
Reactions with 1-Aza-1,3-butadienes

1-Aza-1,3-butadienes [9,11] differ significantly from their oxadienic analoga since they may react as electron-rich as well as electron-deficient dienic component in aza Diels-Alder reactions. Calculations using ab initio and semiempirical methods concerning the behaviour of (*E*)- and (*Z*)-1-aza-1,3-butadiene in such cycloadditions have been recently presented by our group and clearly reveal the tendency of this electronically neutral 1-aza-1,3-butadiene to undergo a two-step cycloaddition mechanism [215].

The somewhat neutral electronic properties of unactivated 1-aza-1,3-butadienes are responsible for their low reactivity towards dienophiles which requires drastic reaction conditions [216]. Another drawback is the inherent instability of the cyclic enamines resulting from the aza Diels-Alder reaction [217]. Therefore, 1-aza-1,3-butadienes have only sparingly been employed for a long period in hetero Diels-Alder chemistry. The main approach made to enhance the reactivity of these compounds is altering the electronical properties by introducing suitable electron-donating or electron-withdrawing substituents.

Thus, Ghosez et al. were successful in showing that *N*,*N*-dimethyl hydrazones prepared from α,β-unsaturated aldehydes react smoothly in normal electron demand Diels-Alder reactions with electron-deficient dienophiles [218, 219]. Most of the more recent applications of such 1-aza-1,3-butadienes are directed towards the synthesis of biologically active aromatic alkaloids and azaanthraquinones [220–224]; a current example is the preparation of eupomatidine alkaloids recently published by Kubo and his coworkers. The tricyclic adduct **3-19** resulting from cycloaddition of naphthoquinone **3-17** and hydrazone **3-18** was easily transformed to eupomatidine-2 **3-20** (Fig. 3-6) [225].

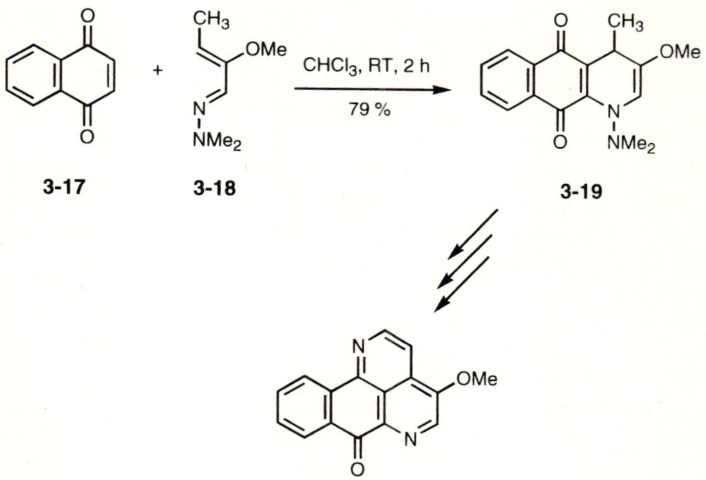

Fig. 3-6

A recent study performed by Ghosez et al. deals with the use of α,β-unsaturated SAMP hydrazones as chiral 1-aza-1,3-butadienes for asymmetric cycloadditions [226]. In this investigation, the reaction of the chiral heterodiene **3-21** with *N*-methylmaleimide afforded the cycloadduct **3-22** in excellent induced diastereoselectivity (Fig. 3-7). Thus, the selectivities obtained are very promising, but the application of this method is restricted to highly reactive electron-deficient dienophiles. The complementary approach, an aza Diels-Alder reaction of an 1-aza-1,3-butadiene with a chiral dienophile, has been investigated by Waldner [227].

Other attempts to carry out normal electron demand aza Diels-Alder reactions with 1-aza-1,3-butadienes base on using *N*-alkoxy [228], *N*-silyloxy [229] and *N*-acylamino-1-aza-1,3-butadienes [230].

In contrast to the hydrazones mentioned above, α,β-unsaturated *N*-sulfonyl imines react as electron-deficient diene component in aza Diels-Alder reactions. In addition to several investigations dealing with their intermolecular cycloadditions under thermal and under high pressure conditions [231–234], Boger's

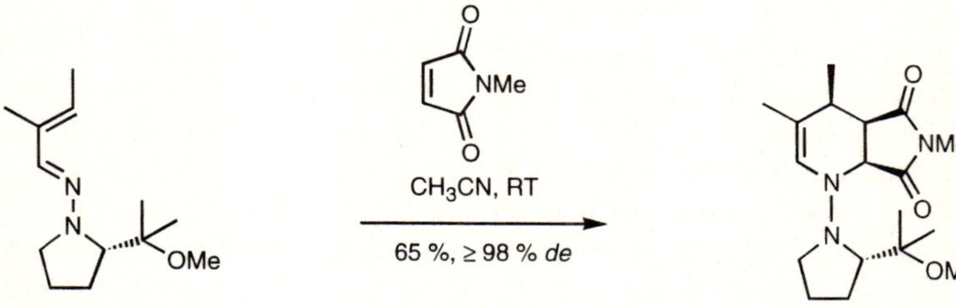

Fig. 3-7

group has as well demonstrated the suitability of these compounds for intramolecular transformations. In this event, upon simply heating a solution of the doubly activated 1-aza-1,3-butadiene **3-23** in toluene, the tricyclic adduct **3-24**, accompanied by minor amounts of aromatisation product **3-25**, was obtained in very good diastereoselectivity [235] (Fig. 3-8).

Aza Diels-Alder reactions of *N*-acyl-1-aza-1,3-butadienes have also been studied and applied in synthesis [236–239]. Interestingly, Jung et al. demonstrated the possibility to generate *N*-acyl-1-aza-1,3-butadienes like **3-27** by flash vacuum pyrolysis of 2-azetines such as **3-26**. Subsequent cycloaddition yielded the bicyclic product **3-28** which is a useful intermediate in alkaloid synthesis [240] (Fig. 3-9).

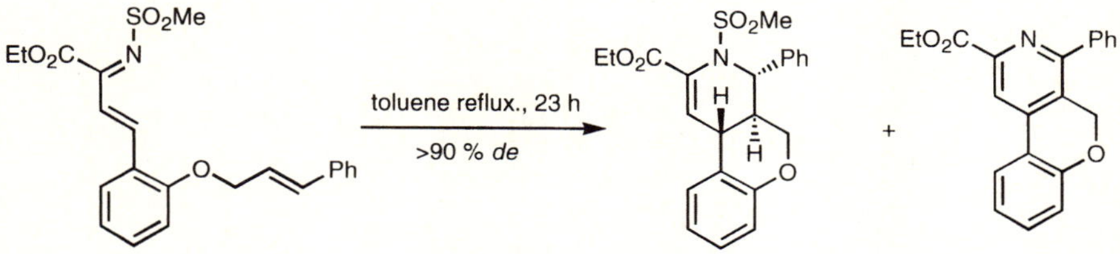

SO_2Me EtO_2C N Ph O
toluene reflux., 23 h
>90 % *de*
3-23 **3-24**, 58 % **3-25**, 0-12 %

Fig. 3-8

N O
FVP
300-540 °C
5-30 mm Hg
benzene, reflux., 28 h
46 % from **3-26**
3-26 **3-27** **3-28**

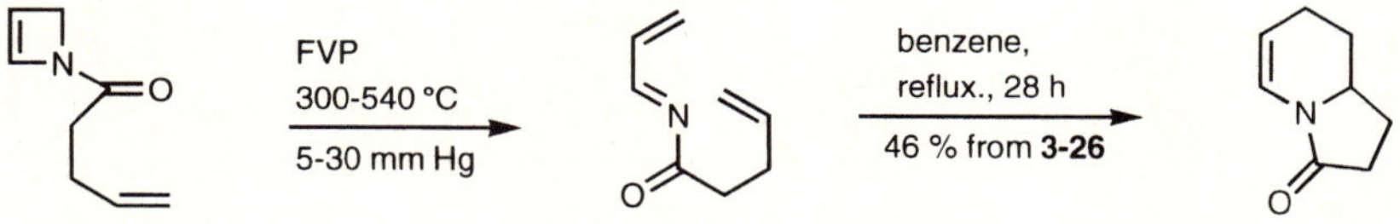

Fig. 3-9

Another elegant application of pericyclic ring opening reactions for an in situ release of 1-aza-1,3-butadienes has been worked out by Wojciechowski by extrusion of sulfur dioxide from benzosultams [241, 242].

The decrease of electron density necessary for inverse electron demand aza Diels-Alder reactions can also be effected by appropiate substituents attached to C-2, especially by a cyano moiety. Several 1-aza-2-cyano-1,3-butadienes have been investigated by Fowler and his coworkers in this context [243–245]. Strikingly, the easily accessable 1-aza-1,3-butadiene **3-29** undergoes cycloaddition not only with electron-rich dienophiles, it reacts as well with neutral and even with electron-deficient dienophiles (Fig. 3-10) [246, 247].

Another novel, highly reactive 1-aza-1,3-butadiene which is derived from benzothiazole reacting with several dienophiles at room temperature in the absence of any catalyst has been described by Sakamoto et al. very recently [248].

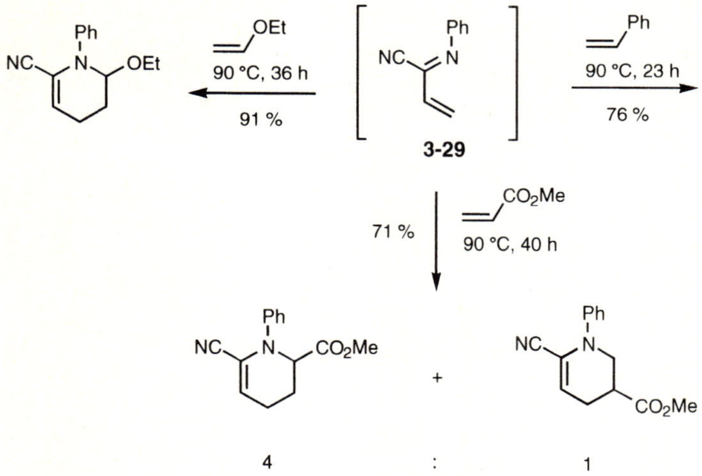

Fig. 3-10

However, activation with suitable substituents is not the only method to make 1-aza-1,3-butadienes react in aza Diels-Alder reactions. Nevertheless, Lewis acid catalysis has not yielded good results in this area of hetero Diels-Alder chemistry. Interestingly, Lewis and Brønsted acids promote a [3 + 2] cycloaddition process instead of the aza Diels-Alder reaction upon addition of α,β-unsaturated hydrazones to quinones [249].

Blechert et al. have opened a very elegant alternative to the classical hetero Diels-Alder methodology by generating the cationic radical **3-31** - which might be conceived as 1-aza-1,3-butadiene - from the 2-vinyl indole **3-30** by means of a single electron transfer process. In the presence of the tetrahydropyridine derivative **3-32** the radicalic intermediate was transformed into the tetracyclic compound **3-33** which contains the complete skeleton of the alkaloid goniomitine [250, 251]. The authors have postulated a stepwise mechanism for this [4+2] process, therefore it should be understood as a formal hetero Diels-Alder reaction (Fig. 3-11).

Considerable acceleration of cycloadditions involving electron-rich 1-aza-1,3-butadienes by ultrasound irradiation has been observed recently [252, 253]. Thus, the application of sonochemical techniques might emerge as helpful tool for Diels-Alder reactions of 1-aza-1, 3-butadienes.

3.3 Reactions with 2-Aza-1, 3-butadienes

Like 1-aza-1, 3-butadienes, 2-aza-1, 3-butadienes may react as electron-rich or as electron-deficient component in aza Diels-Alder reactions upon appropriate substitution; however, they differ in not necessarily needing an activation by appropriate substituents for sufficiently high reactivities. Another significant difference to the reactions of 1-aza-1,3-butadienes is the widespread use of

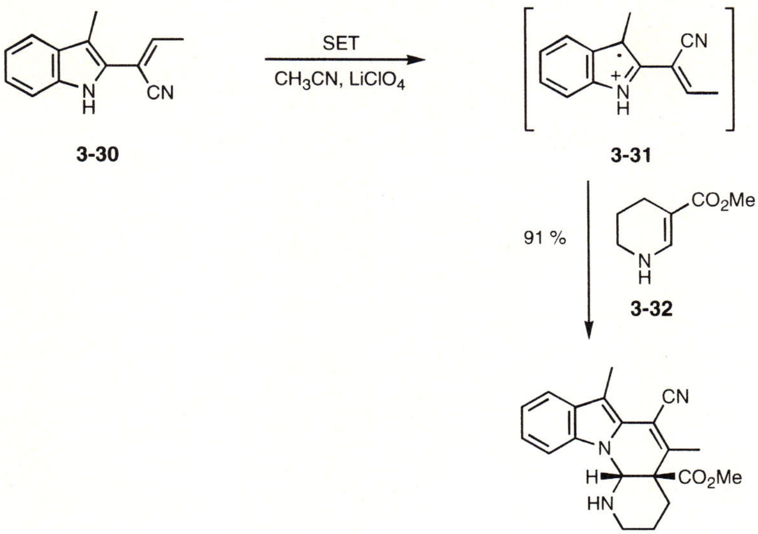

SET
CH_3CN, $LiClO_4$

3-30 3-31

91 %

3-32

3-33

Fig. 3-11

Lewis acid catalysis for aza Diels-Alder reactions of 2-aza-1,3-butadienes [254]. In analogy to the hitherto discussed aza Diels-Alder reactions, evidence for a non-concerted mechanism of these transformations has emerged. Thus, Mellor et al. have found that under suitable conditions azaanthraquinone **3-34** does not only form the expected cycloadduct **3-37** upon treatment with α-methylstyrene and formaldehyde, but the tertiary alcohol **3-36** is also generated presumably via cation **3-35**. Alcohol **3-36** is easily converted into the cycloadduct **3-37** and **3-35** is therefore supposed to act as intermediate in a non-concerted multistep sequence (Fig. 3-12) [255, 256]. More recent studies on *N*-arylimines performed by Laschat et al. have corrobated the assumption that non-concerted processes represent a noteworthy mechanistic pathway in reactions of 2-aza-1,3-butadienes with suitable dienophiles [257].

The reaction sequence depicted in Fig. 3-12 involves an in situ generation of a 2-aza-1,3-butadiene and thus represents a typical domino process [3, 4]. It has found an interesting application in the synthesis of aza steroids [258, 259]. This elegant approach takes advantage of Grieco's observation that in reactions of *N*-aryl imines with cyclopentadiene the latter compound is employed exclusively as dienophilic (!) component [260].

The reactivity of 2-aza-1,3-butadienes lacking any activation by electron density influencing substituents in hetero Diels-Alder reactions with carbo and hetero dienophiles has been thoroughly studied by Barluenga et al. [261]. As valuable application of this methodology a stereoselective approach to 1,3-amino alcohols has been developed [262, 263]. The generation of chiral, halogenated 2-aza-1,3-butadienes like **3-38** allowed to investigate the diastereofacial selectivity in cycloaddition reactions with different dienophiles, and upon treatment

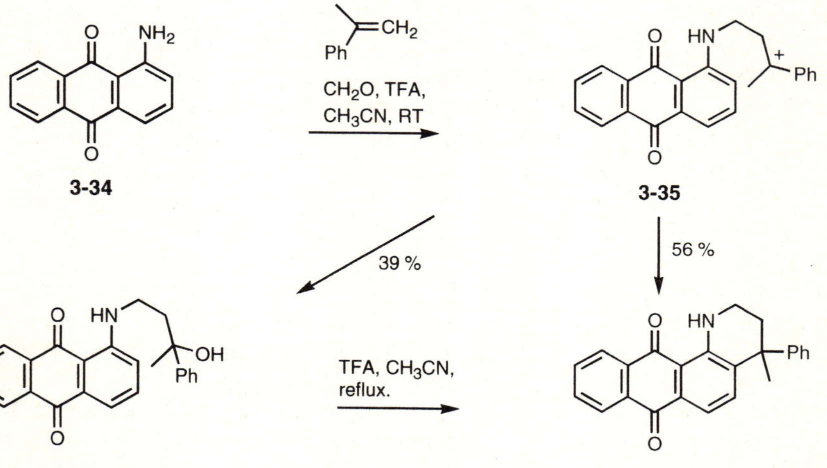

Fig. 3-12

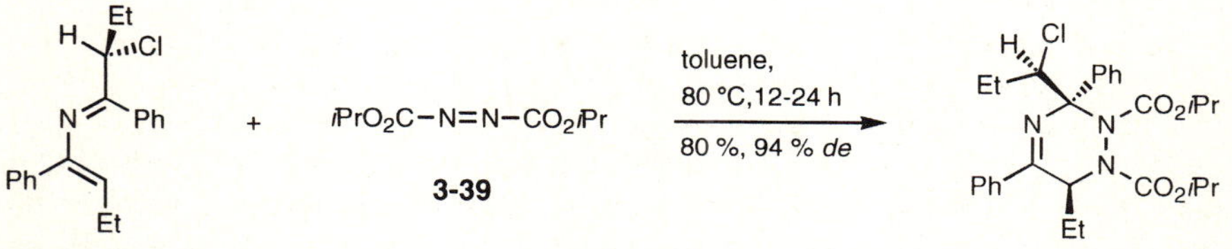

Fig. 3-13

of **3-38** with the azo compound **3-39**, the 1,2,4-triazine **3-40** was formed in very good diastereoselectivity (Fig. 3-13) [264, 265]. Similar cycloadducts derived from **3-38** type 2-aza-1,3-butadienes have turned out to be easily convertable into 1*H*-1,4-diazepine derivatives [266, 267].

An intramolecular aza Diels-Alder reaction of as well electronically neutral *N*-aryl imines useful for the synthesis of novel tetrahydropyridine derivatives has been introduced by our group [268]. The reactive intermediate **3-43** exhibiting the 2-aza-1,3-butadiene subunit was generated in situ from the aldehyde **3-41** and the amino isoxazole **3-42** and led directly to the diastereomerically pure cycloadduct **3-44** (Fig. 3-14). In contrast to the reactions studied by Barluenga, the 2-aza-1,3-butadiene acts as electron-deficient component in this case.

Strikingly, some changes regarding the substitution pattern dramatically influenced the stereoselectivity of this hetero Diels-Alder reaction. Upon attachment of two methyl groups to the dienophilic terminus, the stereoselectivity was almost lost entirely, and additional substitution of the benzene moiety with two chlorine atoms at C-2 and C-4 resulted in a complete reversal of stereoselectivity.

Intramolecular cycloadditions of *N*-aryl imines [269] have also found widespread use in the synthesis of tri- and tetracyclic compounds like octahydroacridine derivatives [270–273]. In these studies tricarbonylchromium comple-

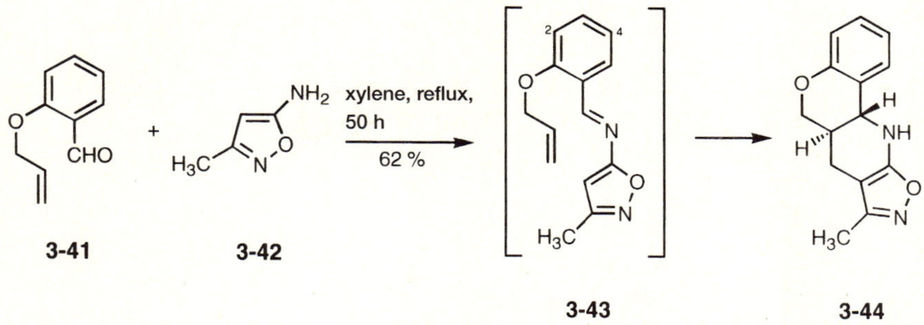

Fig. 3-14

xes have been employed in order to control the stereochemical course. Thus, imino complex **3-45** was smoothly converted into the cycloadduct **3-46** which upon oxidative demetalation yielded the octahydroacridine derivative **3-47** as single diastereomer (Fig. 3-15) [274]. However, this investigation which was carried out by Laschat et al. does not include the use of enantiomerically pure chromium complexes.

An approach directed towards the synthesis of more complex octahydroacridines has been described by Beifuss and his coworkers very recently. Starting from aniline **3-48**, the reactive *N*-aryl iminium ions **3-50** and **3-51** were generated by reaction with a diastereomeric mixture of the aldehyde **3-49**. These intermediates underwent a cationic $[4^+ + 2]$ cycloaddition via an *exo-E-anti* transition state to give the octahydroacridines **3-52** and **3-53** as only products out of 16 possible stereoisomers (Fig. 3-16). The *cis*/*trans*-ratio as well as the (*E*)-configuration of the dienophilic double bond were completely retained during the

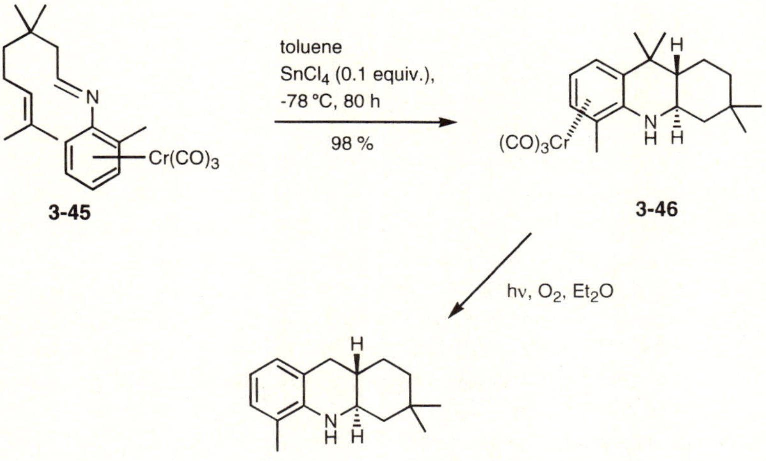

Fig. 3-15

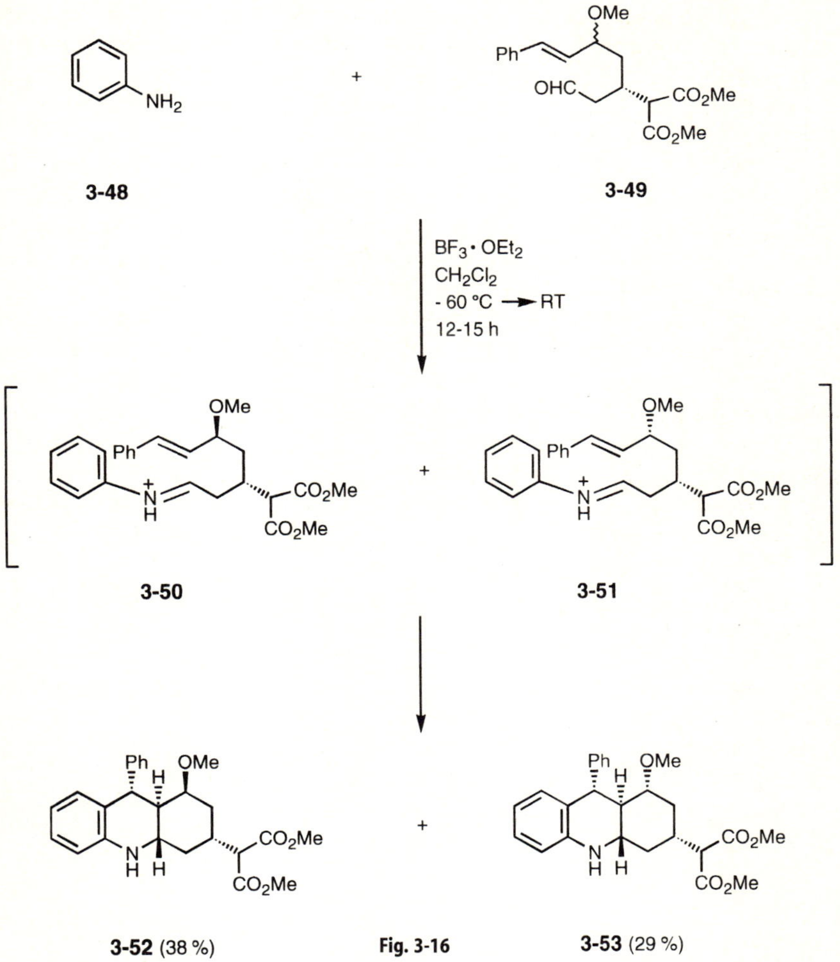

Fig. 3-16

whole domino process [275]. The described methodology basing on cationic 2-aza-1,3-butadienes takes advantage of the enhanced reactivity of iminium ions and has successfully been applied to the stereoselective preparation of 1,2,3,4-tetrahydroquinolines [276, 277].

Activation of 2-aza-1,3-butadienes for inverse electron demand aza Diels-Alder reactions can also be achieved by introducing electron-withdrawing substituents. Thus, Barluenga's group has developed 3,4-bismethoxycarbonyl-2-aza-1,3-butadienes which undergo smooth intramolecular cycloadditions upon heating [278].

On the other hand, also the reverse activation of 2-aza-1,3-butadienes with electron-donating substituents has intensively been studied. Ghosez et al. have originally developed this synthetical tool for the preparation of pyridones, piperidones and pyrimidones [279–281]. More recently, this group extended the scope of this reaction to the asymmetric α-functionalisation of carboxylic acids. In this example, 2-aza-1,3-butadiene **3-54** underwent a stereoselective cycloaddition with the chiral nitroso dienophile **3-55**. The cycloadduct **3-56** was reductively cleaved to give **3-58** which upon hydrolysis yielded the desired amino acid **3-57** in enantiomerically pure form (Fig. 3-17) [282].

Ghosez et al. could also achieve high asymmetric inductions by reacting electron-rich 2-aza-1,3-butadienes with α,β-unsaturated chiral oxazolines [283]. Other applications of electron-rich 2-aza-1,3-butadienes in normal electron demand aza Diels-Alder reactions have been aimed at the preparation of natural cibrostatines [284] and azaanthraquinones [285].

A less usual type of 2-aza-1,3-butadienes is the class of C=C-conjugated carbodiimides like **3-59**. They are readily available from iminophosphoranes and react smoothly with carbo- and hetero dienophiles to yield the desired heterocycles, e.g. 1,3-oxazine **3-60** (Fig. 3-18) [286–288].

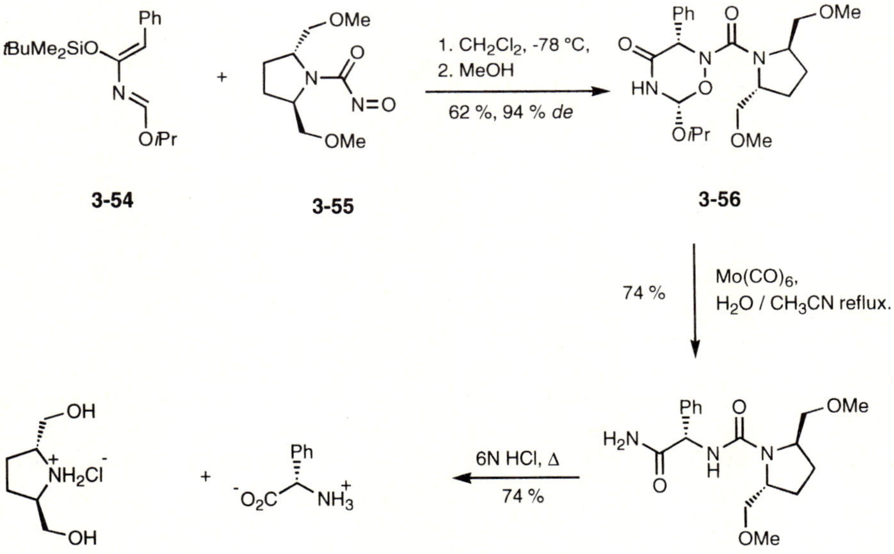

3-54 **3-55** **3-56** **3-57** **3-58**

Fig. 3-17

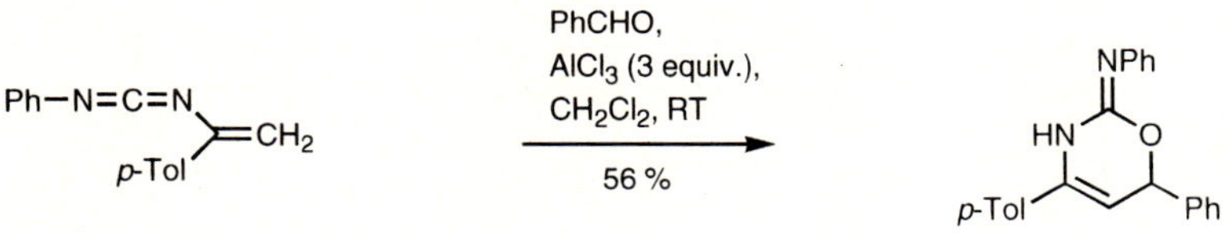

3-59 **Fig. 3-18** **3-60**

3.4 Reactions with N=N Dienophiles

Azo compounds like esters or imides of azo dicarboxylic acid act as reactive dienophiles in normal electron demand hetero Diels-Alder reactions due to the strong activation caused by two electron-withdrawing moieties. In the last years, considerable attention has focused on alkyl and phenyl derivatives of 1,2,4-triazoline-3,5-diones since their cycloadditions to chiral dienes proceed with often excellent facial selectivities. Thus, when reacting an oxapropellane derived diene with *N*-methyltriazolinedione, Paquette et al. obtained the cycloadduct as single diastereomer, but both maleic anhydride and *N*-phenyl maleimide were distinctly less reactive and turned out to undergo cycloadditions with poor selectivities [289].

There exist numerous studies dealing with hetero Diels-Alder reactions of triazolinediones with chiral dienes. In these investigations, Barluenga et al. achieved high stereoselectivities in cycloadditions of *N*-phenyltriazolinedione with chiral 2-alkoxy-1,3-butadienes [290] and Enders et al. were successful in using chiral 2-amino-1,3-butadienes bearing a C_2 symmetrical morpholino moiety in a similar study [291]. The use of dienes with a chiral substituent attached to C-1 has been described by Franck's [292] and Breitmeier's groups [293], and the successful use of a novel, highly oxygenated diene has enlarged the scope of these transformations [294]. Recently, Stoodley et al. have introduced the sugar-linked chiral diene **3-61** which reacts highly diastereoselectively not only with triazolinediones, but also with the acyclic azo dienophile **3-62** to give the pyridazine derivative **3-63** (Fig. 3-19) [295, 296]. Again, the hetero dienophiles turned out to be superior to *N*-phenylmaleimide or tetracyanoethylene with regard to facial selectivity.

Furthermore, azo dienophiles have been employed in diene-transmissive hetero Diels-Alder reactions of cross-conjugated trienes which allow the straightforward construction of polycyclic compounds [297]. Theoretical interest has been directed to the hetero Diels-Alder reaction of diethyl azo dicarboxylate with 1,3-cyclohexadiene whose concerted course was demonstrated by means of a high pressure study [298].

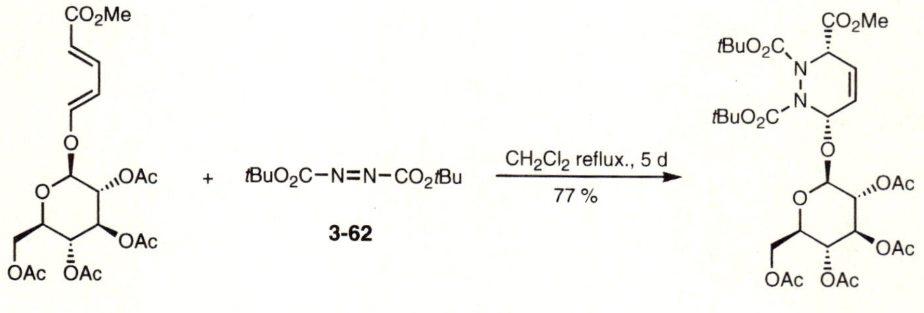

CO_2Me OAc OAc OAc OAc O O

+ $tBuO_2C{-}N{=}N{-}CO_2tBu$

CH_2Cl_2 reflux., 5 d

77 %

3-61 **3-62**

CO_2Me $tBuO_2C$ N N $tBuO_2C$ O O OAc OAc OAc OAc

3-63
single diastereomer

Fig. 3-19

3.5 Reactions with Diaza-1,3-butadienes

A plethora of different acyclic and cyclic diaza dienes has been employed in aza Diels-Alder reactions. With regard to acyclic dienes, the main interest has focused on the cycloadditions of 1,3-diaza-1,3-butadienes. A current example of these transformations is the preparation of highly substituted pyrimidine derivatives such as **3-65** by cycloaddition of diaza-1,3-butadienes e.g. **3-64** with electron-deficient acetylenes (Fig. 3-20) [299].

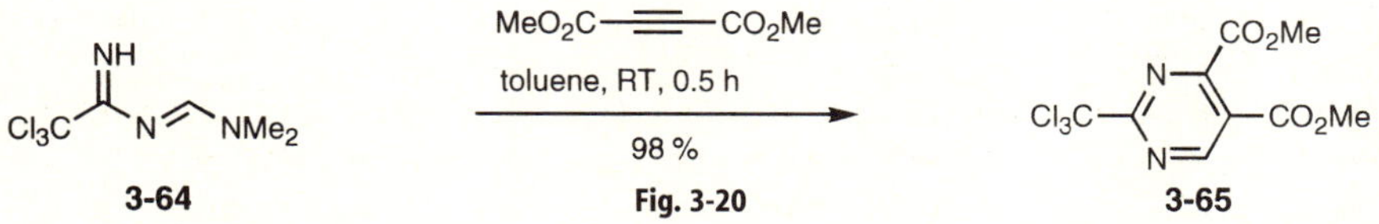

Fig. 3-20

Acyclic 1,3-diaza-1,3-butadienes undergo hetero Diels-Alder reactions not only with ketenes [300–302] and oxazolines [303, 304], they also react with electron-rich dienophiles such as enamines [305, 306]. Intramolecular transformations of this kind efficiently yield complex polycyclic molecules such as the tetracyclic compounds **3-69** and **3-70**. These adducts were synthesised by our group starting from the aldehyde **3-66** and the thiadiazole **3-67**. Under the reaction conditions in situ generation of the isomeric 1,3-diaza-1,3-butadienes **3-68** occurred followed by a hetero Diels-Alder reaction to give the cycloadducts (Fig. 3-21) [307]. Strikingly, the stereochemical result of these transformations does not depend on the configuration of the dienophilic double bond and therefore the cycloaddition is thought to follow a two-step mechanism.

The formation of pyridazines from 1,2-diaza-1,3-butadienes and electron-rich dienophiles has been reported [308]; on the other hand, tetrazine and triazole derivatives have been prepared from these heterodienes and azo esters [309]. Aza Diels-Alder reactions of 1,4-diaza-1,3-butadienes have been employed for the synthesis of unsymmetrical pyrazine derivatives by Heathcock et al. [310].

Cyclic, electron-deficient diaza-1,3-butadienes, e.g. pyrimidines, pyridazines, triazines and tetrazines have proved to be an extremely versatile synthetical tool. Extensive studies aimed at the use of these dienes in the synthesis of natural products stem from Boger's group [11].

Theoretical and synthetical studies carried out by van der Plas et al. deal with intramolecular aza Diels-Alder reactions of ω-alkynylpyrimidines [311, 312]. The substrate **3-71** initially formed a bridged adduct **3-72** upon heating with subsequent release of the fused pyridine derivative **3-73** by retro-Diels-Alder reaction (Fig. 3-22).

Pyridazine dienes behave similarly to pyrimidines inasmuch they as well tend to undergo a cycloaddition-cycloreversion sequence involving the extrusion of nitrogen in the latter step. Condensed pyridazines have been employed

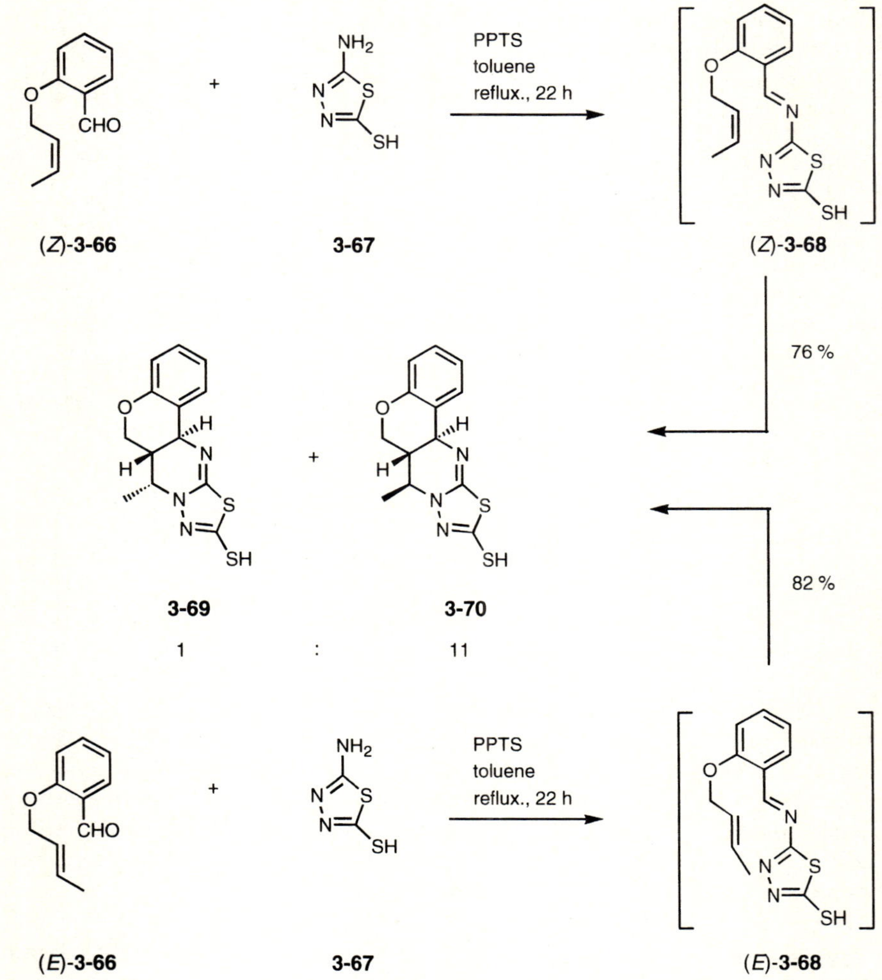

PPTS
toluene
reflux., 22 h

(Z)-3-66 3-67 (Z)-3-68

76 %

3-69 3-70
1 : 11

82 %

PPTS
toluene
reflux., 22 h

(E)-3-66 3-67 (E)-3-68

Fig. 3-21

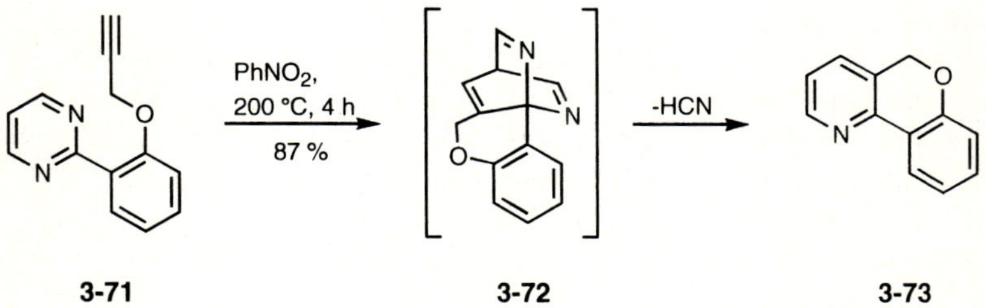

$PhNO_2$,
200 °C, 4 h
87 %

-HCN

3-71 3-72 3-73

Fig. 3-22

as starting materials in the synthesis of isoquinoline derivatives [313, 314], and due to the strong activation effected by two electron-withdrawing substituents, 4,5-dicyanopyridazine **3-74** has emerged as particularly reactive heterodiene [315]. A noteworthy feature of the cycloaddition-cycloreversion sequence mentioned above is the formation of a new diene moiety which is capable to undergo a second cycloaddition. Nesi and Giomi have impressively demonstrated the feasibility of this process by reacting **3-74** with 2,3-dimethylbutadiene which predominantly reacted as dienophilic (!) component [316]. The reaction sequence leading to the tricyclic product **3-76** is thought to proceed via the bicyclic primary adduct **3-75** from which the vinylcyclohexadiene **3-77** is generated upon loss of nitrogen by cycloreversion. A final all-carbon Diels-Alder reaction then yielded **3-76** (Fig. 3-23).

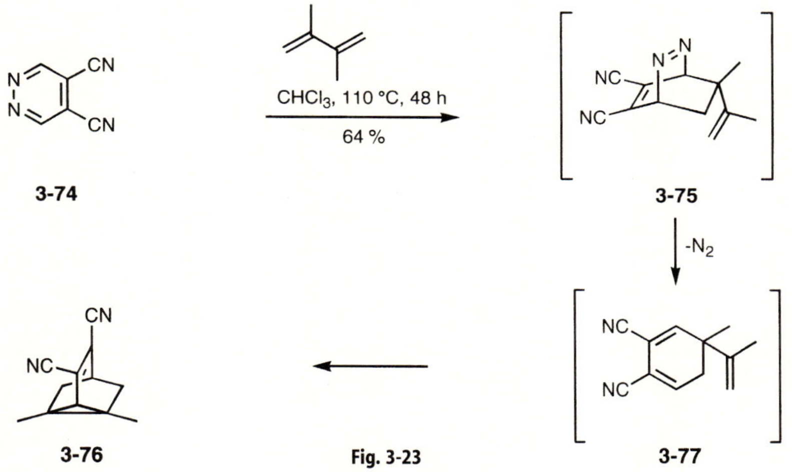

Fig. 3-23

Various triazines have been investigated with regard to their cycloaddition chemistry which again exhibits a pronounced tendency to undergo cycloaddition-cycloreversion sequences. Upon treatment of the electron-deficient 1,3,5-triazine **3-78** with amidine **3-80**, formation of the adduct **3-82** by cycloaddition trapping of the tautomeric enamine **3-81** occurred. Subsequent elimination of ammonia led to the amidine **3-83** which then tautomerized to **3-84**. This intermediate underwent a retro-Diels-Alder reaction leading to the highly substituted amino pyrimidine derivative **3-79** (Fig. 3-24) [317]. This impressive domino reaction developed by Boger et al. has been successfully applied in the total synthesis of P-3A [318], bleomycin A_2 [319] and numerous derivatives thereof [320–324].

Further recent work on cycloaddition chemistry of nitrogen heterocycles deals with 1,2,4-triazines. These cyclic dienes undergo a cycloaddition-cycloreversion series as well; in this case, nitrogen is evolved and thus a pyridine derivative is generated as final product. Snyder et al. efficiently constructed the canthine skeleton by heating the indolyl-tethered 1,2,4-triazine **3-85** which yielded the tetracyclic product **3-86** (Fig. 3-25) [325, 326].

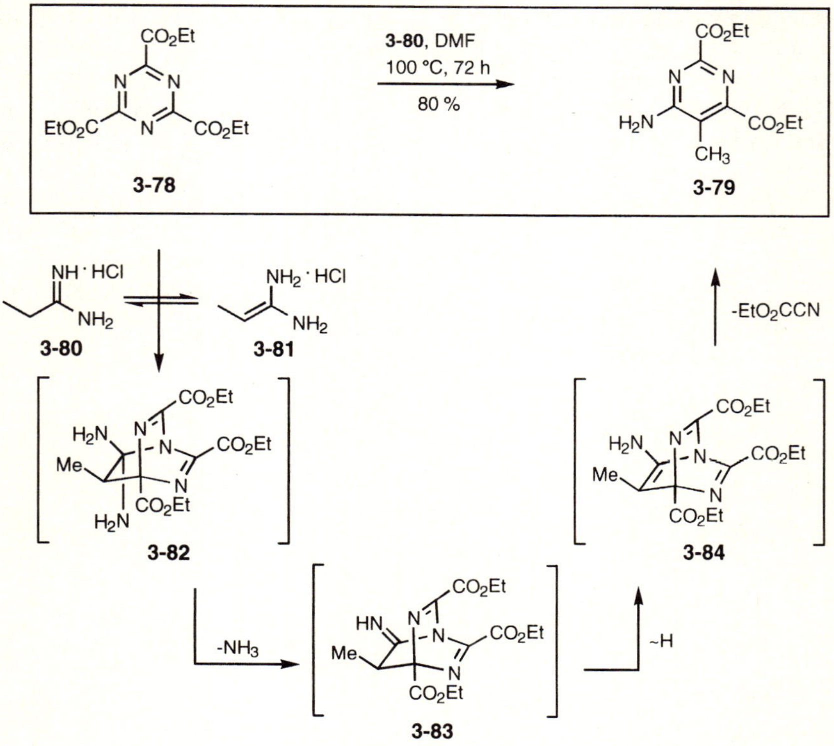

Fig. 3-24

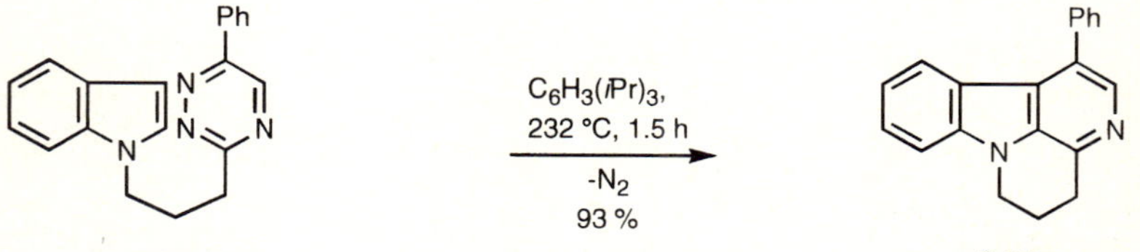

Fig. 3-25

An investigation concerning intramolecular aza Diels-Alder reactions of 3-(ω-alkynyl)-1,2,4-triazines has been published by Taylor et al. [327]; and trichloro-1,2,4-triazine has been introduced as novel triazine diene recently [328]. 1,2,4-Triazines are a useful alternative of 1,4-diaza-1,3-butadienes with regard to the aforementioned synthesis of pyrazines since Taylor's group has found them to undergo cycloadditions with nitriles followed by extrusion of nitrogen [329]. This reaction is noteworthy since it is a Diels-Alder reaction of both electron-deficient diene and dienophile.

The synthesis of pyrrole and indole [330] derivatives by aza Diels-Alder reactions of appropriate 1,2,4,5-tetrazines is another valuable synthetical option based on cycloaddition reactions of cyclic diaza-1,3-butadienes. Using this methodology, Boger and his coworkers efficiently constructed a bipyrrolic intermediate for the synthesis of the DNA cross-linking agent isochrysohermidin [331]. In the presence of excessive 1,2,4,5-tetrazine **3-88**, the diene **3-87** reacted as bisdienophile in a double Diels-Alder reaction followed by the extrusion of nitrogen and methanol to yield the 4,4′-bis-1,2-diazine **3-89**. Subsequent reductive ring contraction then gave the desired bipyrrole **3-90** (Fig. 3-26).

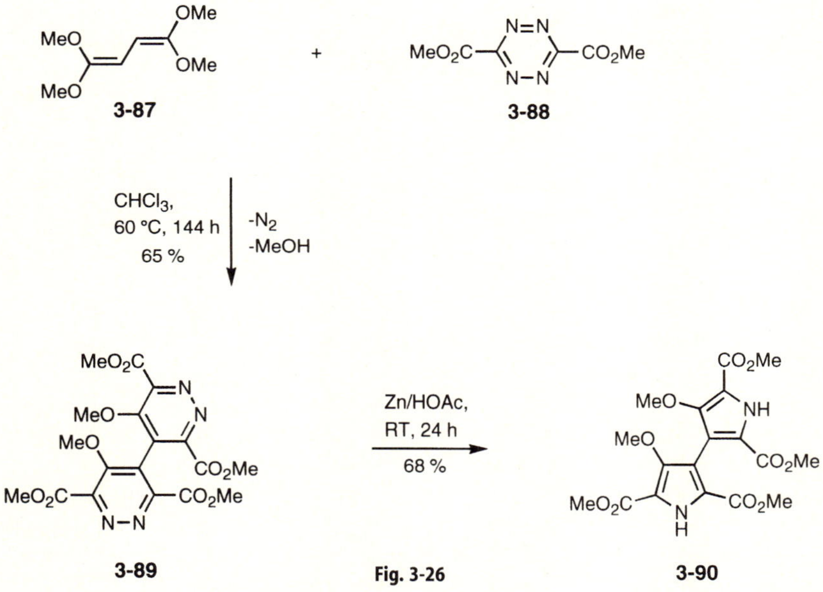

3-87 + **3-88**: $CHCl_3$, 60 °C, 144 h, 65 %; $-N_2$, -MeOH → **3-89**

3-89 → Zn/HOAc, RT, 24 h, 68 % → **3-90**

Fig. 3-26

4 Nitroso- and Nitro Diels-Alder Reactions

4.1 Reactions with N=O Dienophiles

The generation of an 1,2-oxazine **4-1** by hetero Diels-Alder reaction is a transformation which opens a versatile array of highly functionalised acyclic and cyclic structures. Thus, pyrrolidine derivatives **4-2**, amino alcohols **4-3** and aza sugars **4-4** are easily available from these cycloadducts. Cyclic dienes, e.g. **4-5** are converted into bicyclic adducts **4-6** representing straightforward intermediates for aminocyclitols **4-7** (Fig. 4-1).

Since numerous applications of nitroso dienophiles in natural product synthesis [8] and, in particular, aza sugars [332] have been reviewed in the last years, these topics will only be discussed briefly here and only some exemplary and very recent transformations are presented in this article.

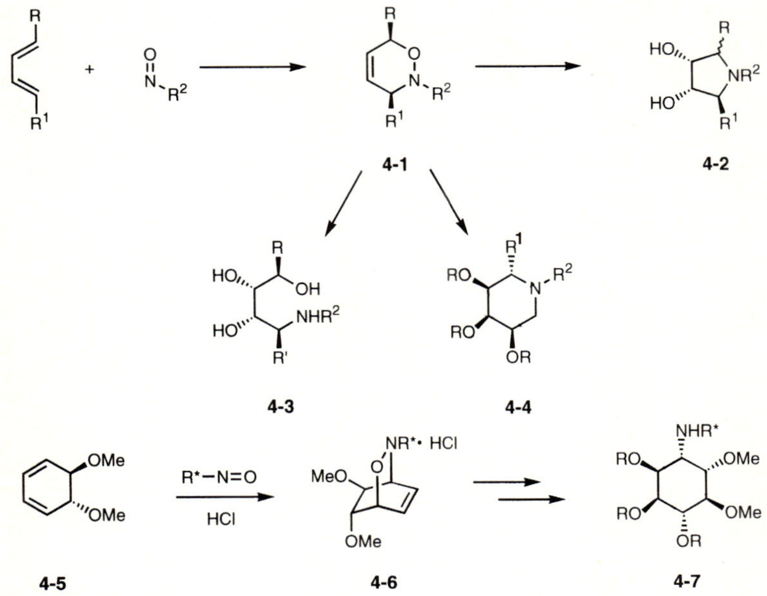

Fig. 4-1

Lewis acid catalysis, apparently dispensible due to the very high reactivity of nitroso dienophiles, has not yet been investigated although such a study has been suggested by Streith and Defoin [8]. Thus, examples of asymmetric catalysis lack completely in this area of hetero Diels-Alder chemistry. Nevertheless, cycloadditions involving nitroso dienophiles have reached an advanced level concerning stereoselectivity and therefore much attention has been paid towards the preparation and application of chiral, enantiopure dienophiles and dienes for these reactions.

The use of the chiral, pyrrolidine derived nitroso dienophile **4-8** for the enantioselective synthesis of amino acids [333] has already been outlined in chapter 3.3, and a multitude of other pyrrolidine derived chiral nitroso dienophiles has been investigated by Streith et al. [334, 335]. Kresze and Vasella have developed sugar derived nitroso dienophiles **4-9** and **4-10** which bear a stereogenic center directly attached to the reactive moiety [336, 337]. Such compounds have proven their synthetical potential in syntheses of aminocyclitols **4-7** [338–340] and azasugars, namely 5-amino-D-allose derivatives [341]. It is worth mentioning that up to four stereogenic centers have been constructed in one step by use of this methodology [338]. Chiral α-hydroxy acyl nitroso dienophiles such as **4-11** have also been found to give good stereoselectivities in cycloaddition reactions. It is thought that a fixed conformation caused by an intramolecular hydrogen bond is responsible for this observation [342, 343]. Furthermore, the pyrroline derived dienophile **4-12** recently introduced by Shustov et al. gave a promising asymmetric induction in a preliminary study [344]. Finally, Streith et al. subjected numerous chiral nitroso dienophiles to cycloaddition reactions with the chiral diene **4-13** in order to study double asymmetric induction in nitroso Diels-Alder reactions (Fig. 4-2) [345].

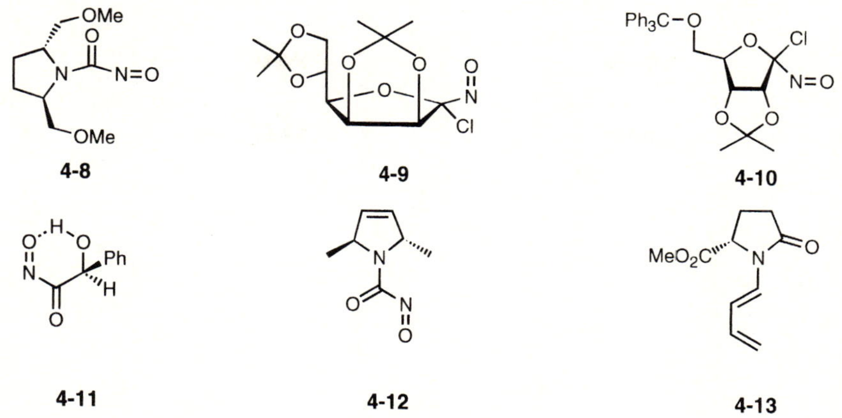

4-8 4-9 4-10

4-11 4-12 4-13

Fig. 4-2

In some recent investigations carried out by Streith's [346, 347] and Wyatt's groups [348], also chiral dienes turned out to exhibit noteworthy potential for asymmetric cycloadditions to achiral nitroso compounds.
However, the often excellent diastereoselectivity is not always accompanied with comparable high regioselection. The following example dealing with structurally more complex substrates might illustrate this problem.
The structurally quite unusal chiral diene **4-14** bearing a β-lactame moiety is known to undergo Diels-Alder reactions with appropriate nitroso dienophiles in excellent yields, but the regioselectivities of such transformations are low [349]. Upon cycloaddition to dienophile **4-15**, the adduct **4-16** was thus formed together with its regioisomer **4-17**. However, both compounds are valuable synthetic intermediates since the main product could be converted into the pyrrolidino sugar derivative **4-18**. The minor regioisomer yielded the erythrose derivative **4-19** [350]. These two compounds represent novel, interesting aza-sugar-β-lactam hybrides (Fig. 4-3).

Another elegant approach directed towards the synthesis of aminoerythrose derivatives such as **4-22** has overcome the regiochemical drawbacks discussed above. Thus, the achiral, pyrrolidine derived diene **4-20** reacted with the nitroso compound **4-15** which was generated in situ from the corresponding hydroxamic acid. Independent of the double bond configuration in the diene, **4-21** was formed as a single diastereomer resulting from a rapid isomerisation of the less reactive (*E*,*Z*)-isomer into the (*E*,*E*)-configurated species. This isomerisation only took place if the periodate used for the in situ generation of the nitroso dienophile was impurified by traces of iodine (Fig. 4-4) [351, 352].

An interesting application of a chiral nitroso dienophile combined with the conversion of the primarily generated 1,2-oxazine into a cyclopentene derivative has been carried out by Miller et al. and is directed towards the synthesis of carbocyclic nucleoside analoga such as **4-28** [353–355]. The alanine derived dienophile **4-23** cycloadded to cyclopentadiene giving two diastereomeric, however easily separable adducts **4-24** and **4-25**. The N–O bond of **4-24** was then reduc-

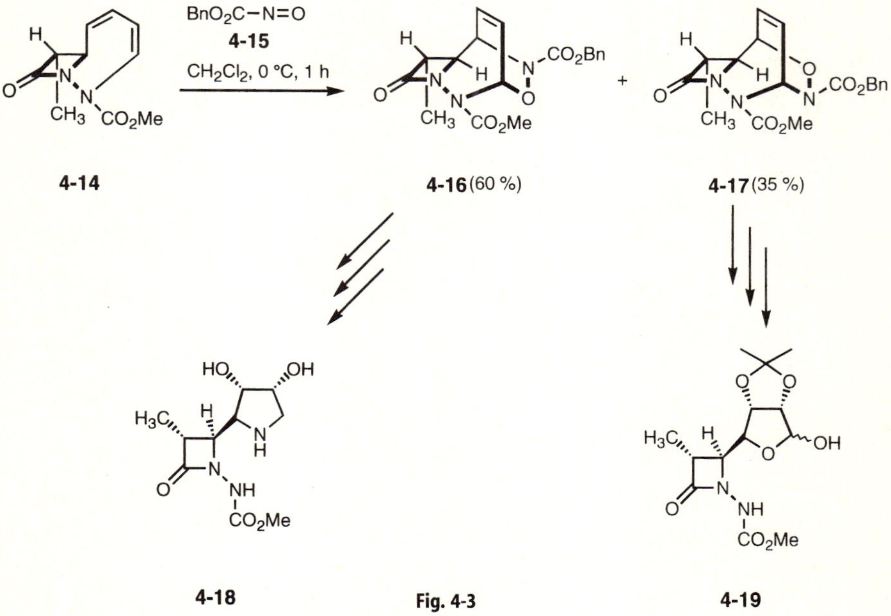

Fig. 4-3

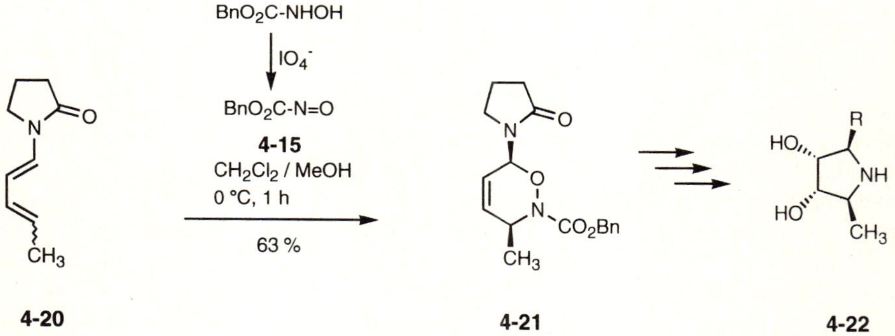

Fig. 4-4

tively cleaved; subsequent acetylation of the resulting intermediate yielded the allylic acetate **4-27** which smoothly underwent a palladium catalysed allylic amination. Again, the main product **4-26** was formed together with some minor regioisomeric byproduct which could be separated. A short sequence then led to **4-28.** Although suffering from some limitations in regio- and diastereoselectivity, this seems to be a novel and innovative application of chiral nitroso dienophiles (Fig. 4-5).

As a current example of a stereoselective intramolecular Diels-Alder reaction using nitroso dienophiles, Kibayashi's studies aimed at the enantioselective total sysnthesis of (–)-pumiliotoxin C **4-31** shall be discussed here [356, 357]. The chiral nitroso compound **4-30** derived from L-malic acid was generated in situ

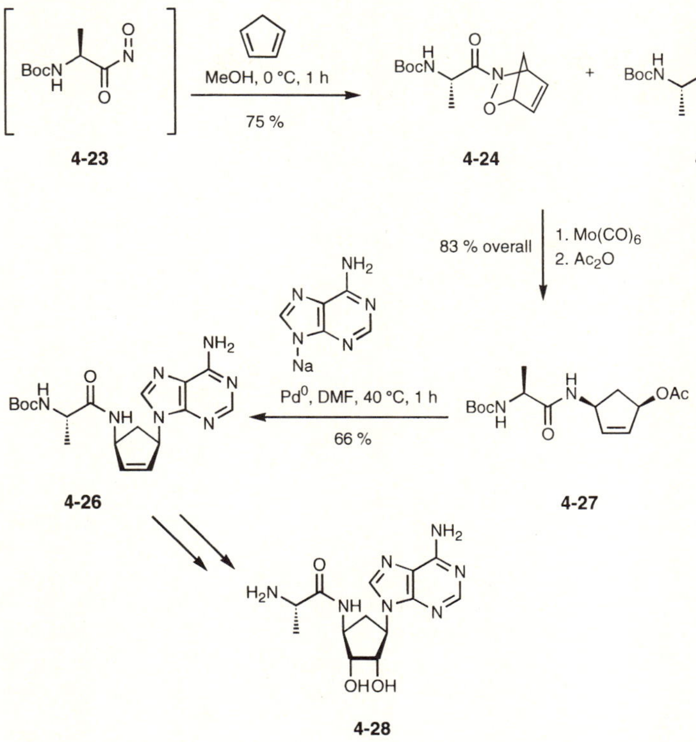

Fig. 4-5

from **4-29** and underwent intramolecular hetero Diels-Alder reaction yielding the bicylic intermediate **4-32** in 4.5:1 diastereoselectivity. This compound proved as suitable for the further conversion into the natural product **4-31** (Fig. 4-6).

Other novel and noteworthy investigations concerning Diels-Alder reactions of nitroso dienophiles are on the one hand directed to novel accesses to these dienophiles [358]; on the other hand, synthetical approaches to nortropanes [359, 360], pyrrolocastanospermine analogs [361] and novel annelated indoles [362] have been developed basing on this powerful preparative tool.

4.2 Reactions with Nitrosoalkenes as Heterodienes

The hetero Diels-Alder reaction of nitrosoalkenes with electron-rich olefins has been known for a long time [363]. A detailed mechanistic study carried out by Reissig et al. has given evidence that this inverse electron demand cycloaddition is a concerted process [364]. Recent ab initio calculations dealing with the reaction between ethylene and nitroso ethylene strongly corrobate this view [365]. In this work, Jursic and Zdravkovski have also investigated the influence of BH_3 as Lewis acid catalyst. However, cycloadditions of nitrosoalkenes already pro-

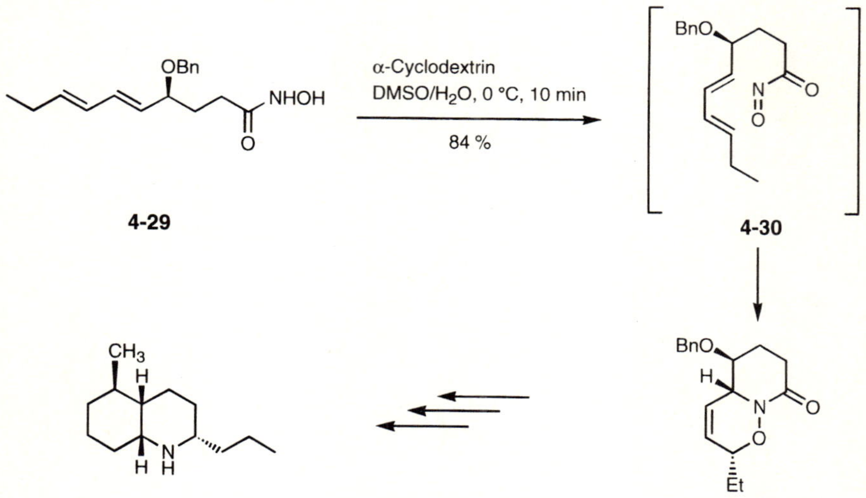

Fig. 4-6

ceed under very mild conditions in the absence of such catalysts; thus, the necessity of using Lewis acids in these transformations is apparently limited.

Figure 4-7 shows a typical hetero Diels-Alder reaction of a nitrosoalkene. Upon in situ generation of the heterodiene **4-34** from the oxime **4-33**, cycloaddition occurred in the presence of the silyl enol ether **4-35** to give the 5,6-dihydro-*4H*-1,2-oxazine **4-36** in excellent yield [366]. Such conversions are very suitable for achieving kinetic resolutions of *E*-/*Z*-isomeric silyl enol ethers since the *Z*-isomers are distinctly less reactive towards **4-34** [367].

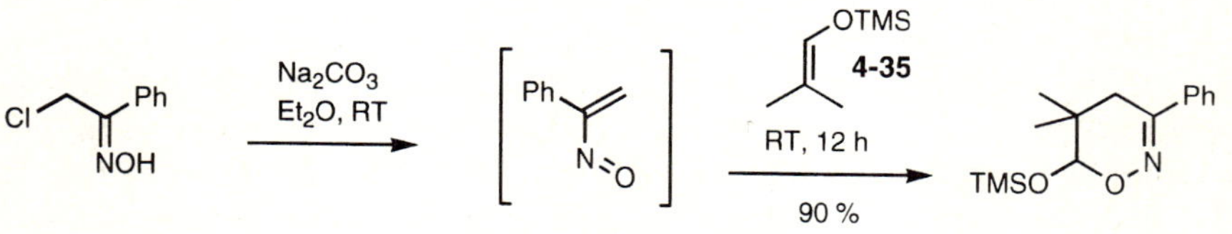

Fig. 4-7

Replacement of the phenyl moiety in **4-34** by a strongly electron-withdrawing substituent leads to a lower electron density and thus enhances the reactivity of the heterodienic compound. Indeed the corresponding trifluoro methyl derivative underwent cycloadditions with alkenes which had failed to react with **4-34** [368]. The activated nitrosoalkene **4-37** has been applied to the synthesis of proline derivatives. Cycloadduct **4-39** resulting from the cycloaddition of in situ generated **4-37** with enamine **4-38** gave the desired proline derivative **4-40** upon reductive ring contraction (Fig. 4-8) [369].

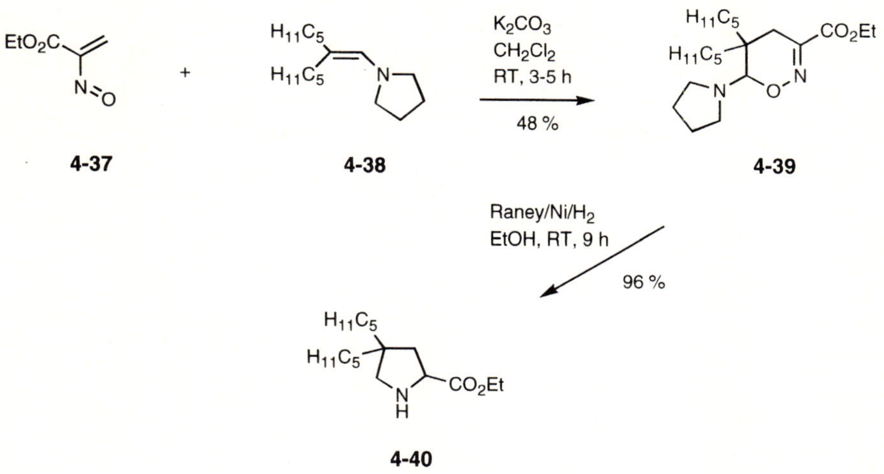

4-37 + 4-38 → (K_2CO_3, CH_2Cl_2, RT, 3-5 h; 48 %) 4-39

4-39 → (Raney/Ni/H_2, EtOH, RT, 9 h; 96 %) 4-40

Fig. 4-8

The array of dienophiles amenable to these hetero Diels-Alder reactions is not limited to enol ethers and enamines since allylsilanes and simple alkenes have also been successfully employed [370, 371]. More recently, it has been shown that methoxy allenes such as **4-41** undergo formation of *6H*-1,2-oxazines **4-43** upon cycloaddition to nitrosoalkenes such as **4-34** and subsequent tautomerisation of the intermediate *exo*-methylene compound **4-42** (Fig. 4-9) [372, 373]. In these studies, **4-43** proved to be a versatile synthetical intermediate allowing oxidative demethylation or reductive removal of the methoxy group as well as nucleophilic substitutions after the generation of an azapyrylium ion [372–374]. Furthermore, ring contraction reactions of these oxazines leading to pyrroles [373] and γ-lactames [375] are known.

Indoles may as well serve as dienophilic compounds for hetero Diels-Alder reactions with nitrosoalkenes. However, the resulting adducts are not stable and undergo further conversion to oximes which represent useful intermediates for the straightforward synthesis of tryptophane derivatives [376, 377].

In order to carry out asymmetric cycloadditions of nitrosoalkenes, Reissig et al. have introduced chiral enol ethers derived from terpenes [378] and from the glucose derivative **4-46** [379]. Using these compounds, considerable asymmetric induction has been obtained; thus, the 5,6-dihydro-*4H*-1,2-oxazine **4-45** was formed by hetero Diels-Alder reaction of **4-34** with chiral **4-44** in good diastereoselectivity (Fig. 4-10) [379].

Other current investigations concerning cycloadditions of nitrosoalkenes are directed towards employment of more complex dienophiles, e. g. *N,N*-bis-trimethylsilyl enamines [380, 381] or 2,5-dihydrooxepines [382]. Furthermore, interest focuses on exploring the scope of subsequent reactions of the cycloadducts, such as stereoselective halogenation [383], *cis*-dihydroxylation [384] and numerous reductive [385] as well as acid or transition metal induced [386] transformations of 5,6-dihydro-*4H*-1,2-oxazines.

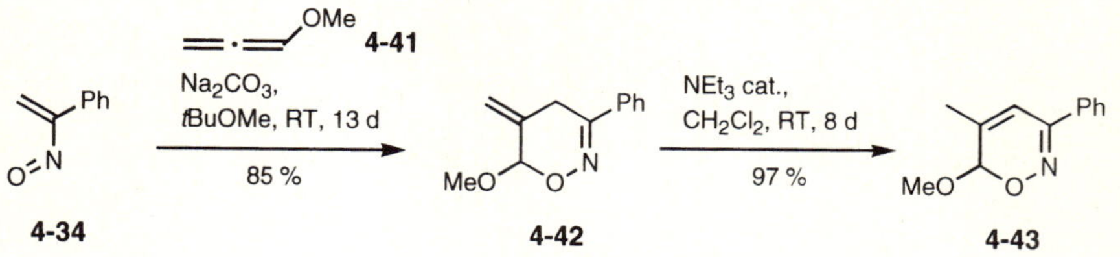

Fig. 4-9

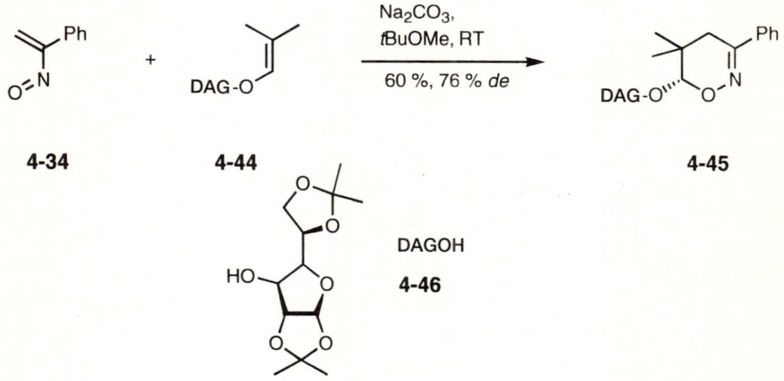

Fig. 4-10

4.3
Reactions with Nitroalkenes as Heterodienes

If nitroalkenes are employed as heterodienes in hetero Diels-Alder reactions instead of nitrosoalkenes, cyclic nitrones are formed. These cycloadducts undergo numerous subsequent reactions, and especially the combination of this hetero Diels-Alder reaction with a 1,3-dipolar cycloaddition is an extremely powerful tool for the synthesis of polycyclic alkaloids. This domino [4+2]/[3+2] cycloaddition chemistry has been comprehensively reviewed by Denmark and Thorarensen very recently, and this review also covers many hetero Diels-Alder reactions of nitroalkenes being not part of this sequential transformation [5]. Therefore the present article will focus on some selected examples which might highlight the advanced state of the art concerning stereocontrol of these reactions. On the other hand, an insight shall be given into the multitude of polycyclic structures accessible by means of nitroalkene cycloaddition chemistry.

Chiral dienophiles, e.g. enol ethers and enamines, allow to conduct these hetero Diels-Alder reactions in a highly stereoselective manner. In an exemplary transformation described by Bäckvall et al. the nitrone **4-50** was formed as a single diastereomer upon treatment of the chiral enamine **4-48** with the nitroalkene **4-49** which was generated in situ from the seleno compound **4-47** [387]. Interestingly, the enamine **4-48** did not only act as dienophile, it also catalysed the initial base induced elimination of PhSeH from **4-47** (Fig. 4-11).

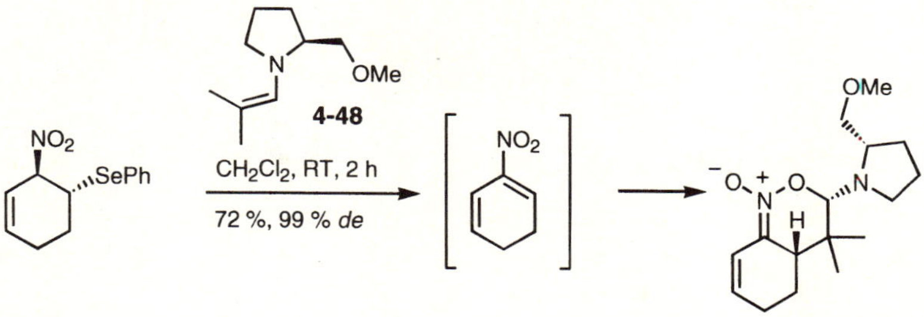

Fig. 4-11

It is as well possible to obtain excellent simple diastereoselectivities in intramolecular cycloadditions of appropriate achiral nitro-1,n-dienes [388].

Cyclic nitrones generated by [4+2]-cycloaddition of nitroalkenes undergo various, synthetically very valuable reactions. Thus, Denmark et al. have developed an elegant access to different enantiopure, 3- and 3,4-substituted pyrrolidine derivatives by reductive ring contraction of the cyclic nitrone resulting from a hetero Diels-Alder reaction [389, 390]. Upon reaction of *E*-2-nitrostyrene **4-51** with the chiral enol ether **4-52** in the presence of the bulky Lewis acid MAPh (**4-53**), three diastereomeric cycloadducts **4-54**, **4-55** and **4-56** were formed. Hydrogenolysis of the main product **4-54** yielded the desired pyrrolidine **4-57** in excellent optical purity and allowed nearly quantitative recovery of the chiral auxiliary (Fig. 4-12) [391]. It is noteworthy that the nature of the Lewis acid catalyst, especially its steric demand, decisively influences the stereochemical course of such cycloadditions [392].

However, far the most powerful synthetical methodology involving cycloaddition chemistry of nitroalkenes is the combination of a hetero Diels-Alder reaction with a 1,3-dipolar cycloaddition of the resulting nitrone. Up to six stereogenic centers may be constructed in the course of this protocol, and a multitude of preparative options results from applying either intra- or intermolecular varieties of the single steps and from the different modes to connect the resulting cyclic entities (Fig. 4-13).

Thus, there exist four subclasses of this sequential transformation, and e. g. the inter [4+2]/intra [3+2] reaction allows the efficient construction of fused and spiro connected tricyclic compounds (Fig. 4-14).

Very recently, Denmark's group was successful in opening a third, bridged cyclisation mode within the inter [4+2]/intra [3+2] subclass of domino [4+2]/[3+2] cycloadditions by attaching the dipolarophilic double bond to the dienophile [393]. The initial Lewis acid catalysed hetero Diels-Alder reaction of the nitroalkene **4-58** and the 1,4-diene **4-59** yielded the nitrone **4-60** as single diastereomer which upon heating smoothly underwent a 1,3-dipolar cycloaddition to give the bridged polycycle **4-61** (Fig. 4-15).

The hitherto discussed transformations clearly demonstrate the great value of using chiral dienophiles for hetero Diels-Alder reactions of nitroalkenes. Recent studies deal with the application of various chiral alcohols in order to get

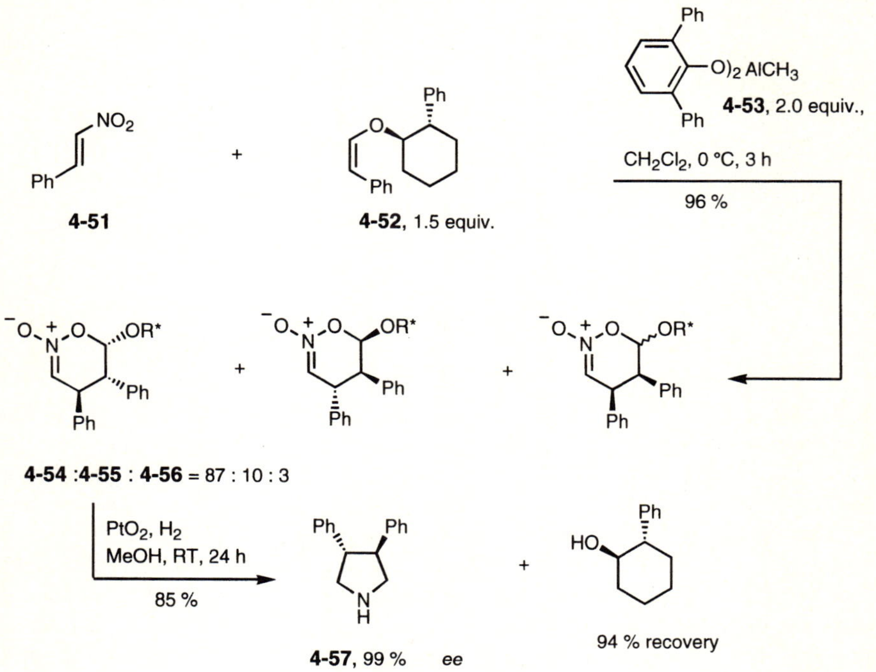

Ph
$O)_2 AlCH_3$
Ph
4-53, 2.0 equiv.,
CH_2Cl_2, 0 °C, 3 h
96 %

NO_2
Ph
4-51
+
Ph
O
Ph
4-52, 1.5 equiv.

OR* Ph Ph + OR* Ph Ph + OR* Ph Ph

4-54 :**4-55** : **4-56** = 87 : 10 : 3

PtO_2, H_2
MeOH, RT, 24 h
85 %

Ph Ph N H
4-57, 99 % *ee*
+
Ph HO
94 % recovery

Fig. 4-12

DPP R^1 R^2 DP
inter [4+2] / inter [3+2]

DPP R^1 R^2 DP
inter [4+2] / intra [3+2]

DPP R^1 R^2 DP
intra [4+2] / inter [3+2]

DPP R^1 R^2 DP
intra [4+2] / intra [3+2]

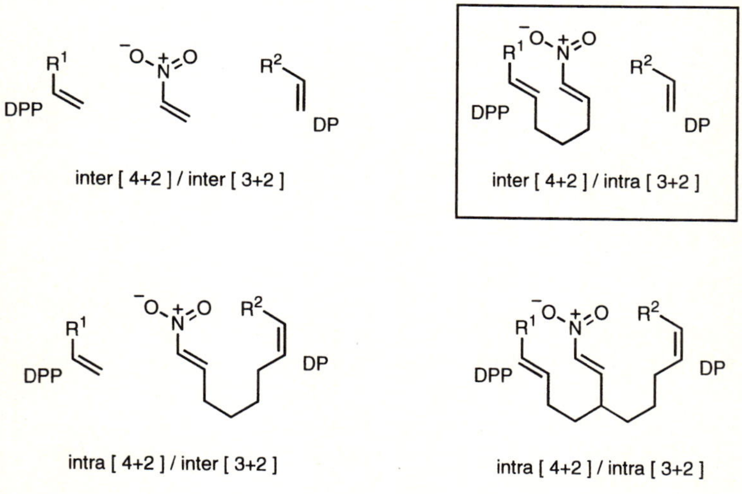

DP = Dienophile, DPP = Dipolarophile

Fig. 4-13

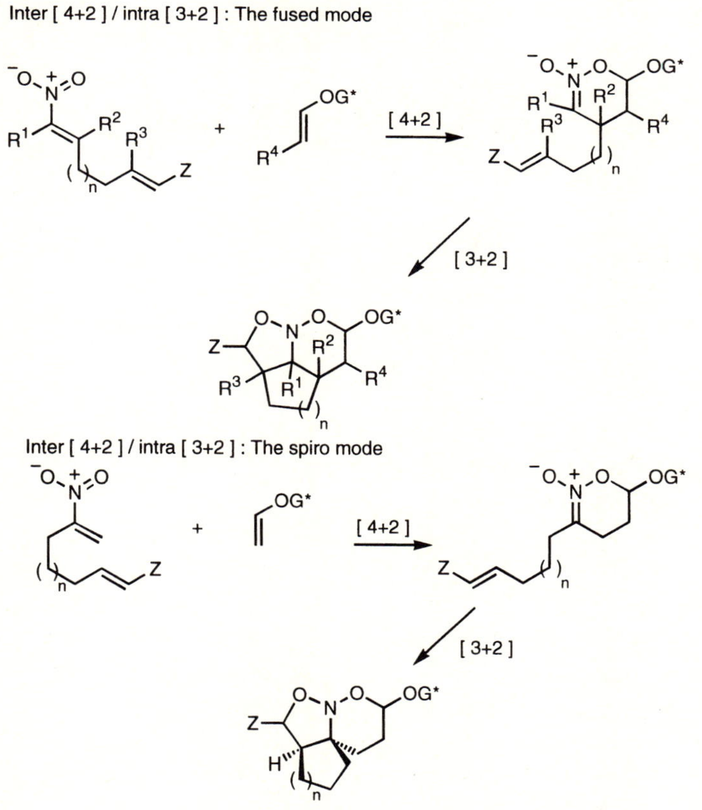

Inter [4+2] / intra [3+2] : The fused mode

[4+2]

[3+2]

Inter [4+2] / intra [3+2] : The spiro mode

[4+2]

[3+2]

Fig. 4-14

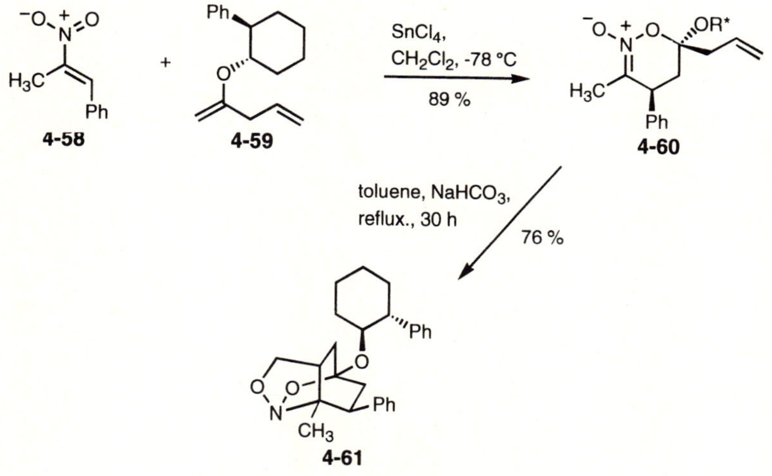

$SnCl_4$, CH_2Cl_2, -78 °C

89 %

4-58 **4-59** **4-60**

toluene, $NaHCO_3$, reflux., 30 h

76 %

4-61

Fig. 4-15

access to novel, highly stereoselectively reacting dienophiles [394]. Chiral nitroalkenes derived from thymidine nucleosides have also attracted considerable attention [395]. The novel carbohydrate derived chiral nitroalkene **4-62** underwent a domino [4 + 2]/[3 + 2] cycloaddition leading firstly to the nitrone **4-63** and then to the bicyclic adduct **4-64** which was obtained as a single diastereomer. Strikingly, this process did not require the presence of any catalyst (Fig. 4-16) [396].

The described sequential transformations involving hetero Diels-Alder reactions of nitroalkenes represent a very mature synthetical tool. Its application to the total synthesis of several alkaloids will be briefly discussed in the Sect. 7 of this article.

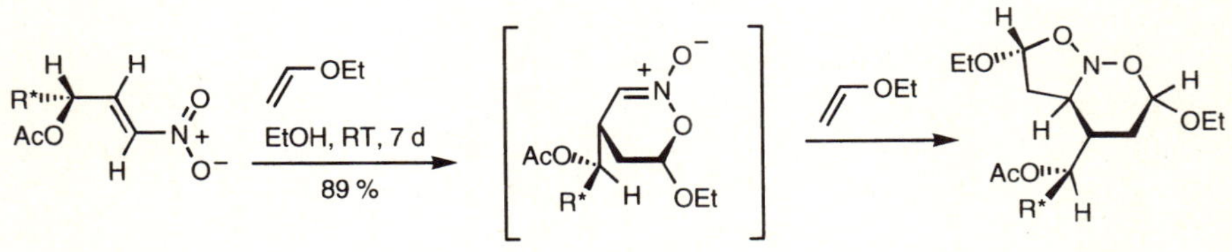

R* = D-*lyxo*-$(CHOAc)_3CH_2OAc$

Fig. 4-16

5 Thia Diels-Alder Reactions

5.1 Reactions with C=S Dienophiles

Thiocarbonyl compounds are well known dienophiles which yield thiopyran derivatives upon hetero Diels-Alder reactions with suitable dienes. Amongst them, the thioaldehydes exhibit the highest reactivity in these cycloadditions, but their low stability often requires to generate them in situ [397, 398]. Interestingly, one of these methods itself represents a retro-thia Diels-Alder reaction since cycloadducts of thioaldehydes and anthracene undergo cycloreversion upon refluxing in toluene [399]. These adducts may be conceived as "chemical stores", and it is even possible to transform the stored thioaldehyde and release it then in modified form by cycloreversion [400]. In addition to hetero Diels-Alder reactions with an external diene, the in situ generated thioaldehyde may undergo ene reactions [401, 402] and, if tethered to a diene, intramolecular cycloadditions [402, 403].

The stereoselectivity of hetero Diels-Alder reactions of thioaldehydes and cyclopentadiene has thoroughly been investigated by Vedejs and his coworkers [404]. In general, these cycloadditions proceed only with moderate *endo* selectivities; for high *endo*-selectivities, very bulky thioaldehydes are necessary. This study includes an investigation of asymmetric induction effected by chiral thioaldehydes, and the racemic heterodienophile **5-2**, generated in situ from **5-1** by photolysis, cycloadded to Danishefsky's diene with excellent facial selectivity to give the enone **5-3** (Fig. 5-1). However, this high level of induced diastereoselection is not yet general in such transformations.

In a more recent study, Koizumi et al. employed terpene derivatives as chiral auxiliaries attached to thioaldehydes [405], but these heterodienophiles gave only moderate diastereoselectivities in reactions with cyclopentadiene. Mazzanti et al. have investigated the hetero Diels-Alder reaction of thioketones bearing an asymmetric silicon atom directly attached to the thiocarbonyl moiety which induced diastereoselectivities up to 50% *de* [406].

Cycloadditions of thiocarbonyl compounds have been employed in the synthesis of biologically active agents. Thus, azocine derivatives [407] and novel opiate antagonists [408] have been prepared using this strategy. More recently, an α-thioketo ester has been transformed into various derivatives of aprikalim which has attracted interest as potassium channel activator [409]. Kirby's group has efficiently constructed the thiashikimic acid derivative **5-7** by releasing the thioglyoxylate **5-5** from its anthracene adduct **5-4** in the presence of the diene **5-6** [410]. The desired thiapyrane **5-7** was easily available from the main cycloadduct **5-8** (Fig. 5-2).

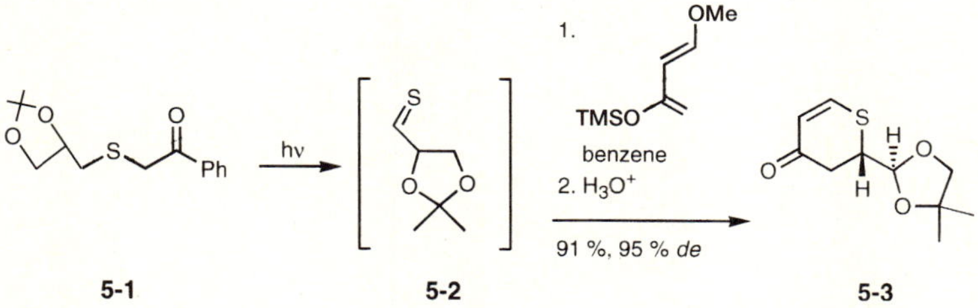

Fig. 5-1

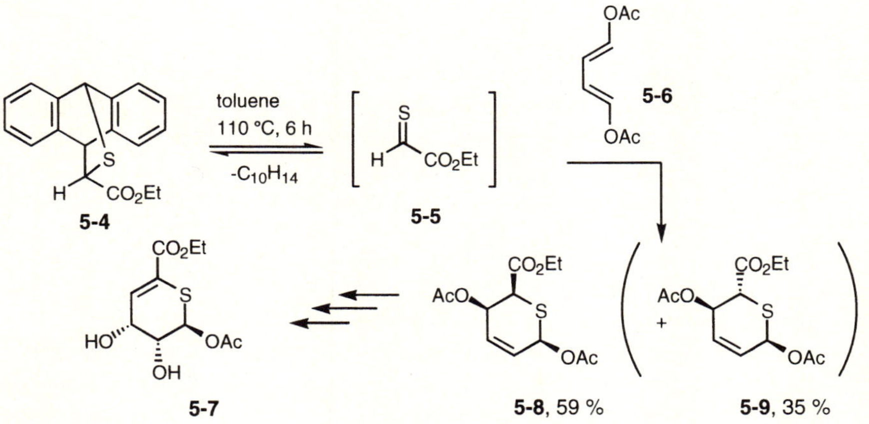

Fig. 5-2

Due to the high reactivity and high sensitivity of thioaldehydes, Lewis acid catalysis has not been applied to thia Diels-Alder reactions. However, Heimgarten et al. clearly demonstrated the suitability of Lewis acids to make less reactive thiocarbonyl compounds, e.g. thiazole thiones, react as heterodienophiles in hetero Diels-Alder reactions [412].

5.2 Reactions with Thia-1,3-butadienes

Similarly to the homologous 1-oxa-1,3-butadienes, 1-thia-1,3-butadienes are known to be very suitable and reactive substrates for hetero Diels-Alder reactions. However, in contrast to the oxa-1,3-butadienes which in general act as electron-deficient component in such cycloadditions, thia-1,3-butadienes predominantly undergo normal electron demand Diels-Alder reactions with electron-deficient dienophiles. Nevertheless, also some reactions of thia-1,3-butadienes involving electron-rich dienophiles have been described [412, 413]. Thia-1,3-butadienes considerably tend to dimerize due to their high reactivity in hetero Diels-Alder reactions [414].

Enaminothiones such as **5-10** bearing an electron-donating dialkylamino moiety have been extensively used as thia-1,3-butadienes [415–417]. In a typical procedure, **5-10** cycloadded to the oxoglutaconic acid derivative **5-11** in the absence of any catalyst under very mild conditions. Subsequent elimination of dimethylamine yields the thiopyran **5-12** (Fig. 5-3) [418].

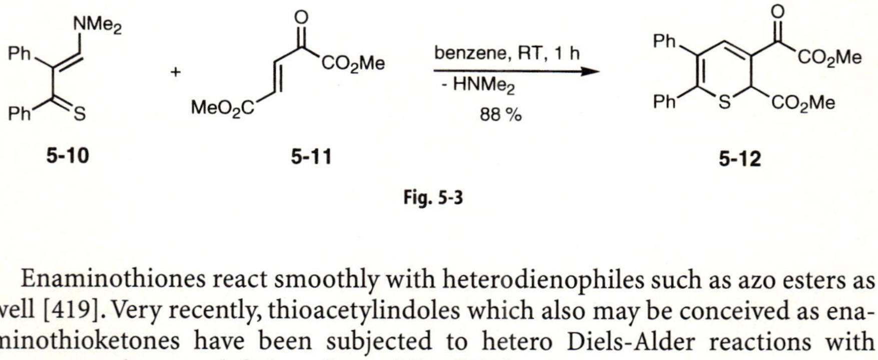

Fig. 5-3

Enaminothiones react smoothly with heterodienophiles such as azo esters as well [419]. Very recently, thioacetylindoles which also may be conceived as enaminothioketones have been subjected to hetero Diels-Alder reactions with numerous electron-deficient dienophiles [420].

As another interesting class of thia-1,3-butadienes, Fishwick et al. have introduced 2-*N*-acylamino-1-thia-1,3-butadienes [421] which strikingly react with electron-rich as well as with electron-deficient dienophiles [422, 423]. This behaviour may be illustrated by the cycloadditions of thia-1,3-butadiene **5-13** with ethyl vinyl ether and acrylonitrile (Fig. 5-4).

1-Thia-1,3-butadienes have been successfully employed in intramolecular hetero Diels-Alder reactions [414, 424, 425]. More recently, some interesting varieties of such intramolecular cycloadditions which allow the efficient construction of sulfur containing polycyles have been worked out by Saito's group.

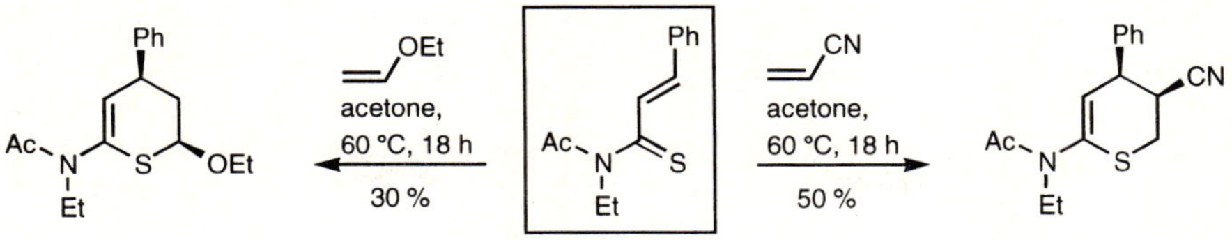

Fig. 5-4

Such syntheses take advantage of transannular intramolecular hetero Diels-Alder reactions [426], another impressive example is the diene transmissive hetero Diels-Alder reaction of the thioketone **5-14**. The diene **5-15** formed by this cycloaddition underwent a second Diels-Alder reaction with *N*-phenylmaleimide to yield the fused polycycle **5-16** as single diastereomer (Fig. 5-5) [427].

The use of Lewis acids in order to catalyze hetero Diels-Alder reactions of thia-1,3-butadienes is not widespread, but recent investigations stemming from Saito et al. reveal a remarkable acceleration of these transformations in the presence of $AlCl_3$ or $EtAlCl_2$ [428]. In a first study concerning asymmetric hetero Diels-Alder reactions of thia-1,3-butadienes, Saito et al. found Lewis acids to have a beneficial effect on the induced diastereoselectivities. Thus, the thioketone **5-17**, generated in situ by thermal cycloreversion from its dimer, underwent a completely *endo*-selective cycloaddition upon treatment with (–)-dimenthyl

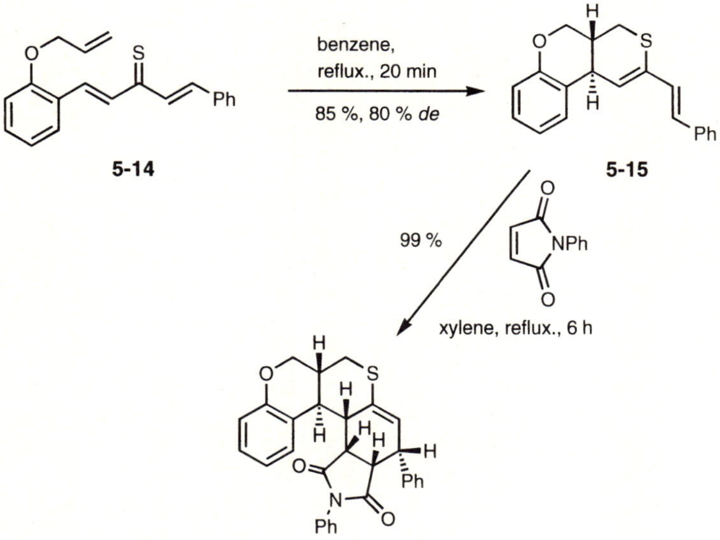

Fig. 5-5

fumarate **5-18** to give the cycloadducts **5-19** and **5-20**. However, the induced diastereoselectivity did not exceed 71% *de* in this study (Fig. 5-6) [429].

Basing on this investigation, **5-17** has been cycloadded to numerous chiral dienophiles in order to increase the asymmetric induction [430]. The use of chiral oxazolidinones allowed to obtain induced diastereoselectivities up to 92% *de*. On the other hand, the complementary application of enantiopure, chiral thia-1,3-butadienes introduced by Fishwick et al. has yielded very promising results [431]. Upon in situ generation of the chiral heterodiene **5-22** from **5-21**, a completely *exo*-selective cycloaddition occurred in the presence of cyclopentene leading to the fused thiopyran **5-23** as a single diastereomer (Fig. 5-7).

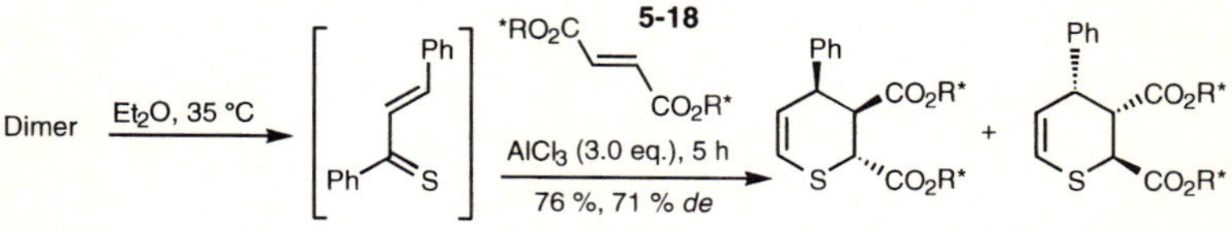

Dimer $\xrightarrow{Et_2O,\ 35\ °C}$ [**5-17**] $\xrightarrow[\text{76 \%, 71 \% } de]{^{*}RO_2C\text{-CH=CH-}CO_2R^{*}\ \textbf{5-18},\ AlCl_3\ (3.0\ eq.),\ 5\ h}$ **5-19** (major) + **5-20** (minor)

R* = (-)-menthyl

Fig. 5-6

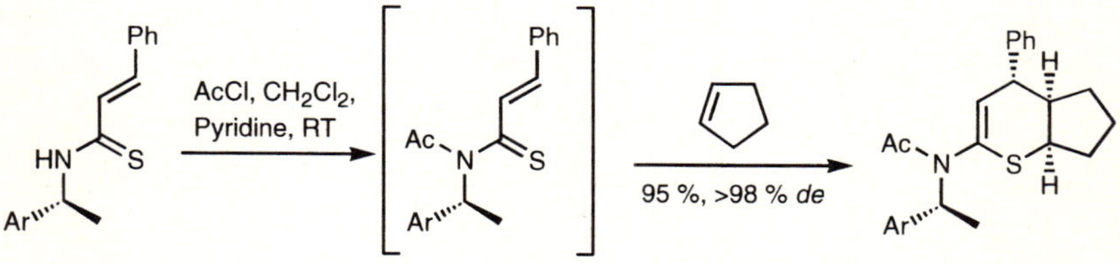

5-21 $\xrightarrow{AcCl,\ CH_2Cl_2,\ Pyridine,\ RT}$ [**5-22**] $\xrightarrow[\text{95 \%, >98 \% } de]{\text{cyclopentene}}$ **5-23**

Ar = 1-naphthyl

Fig. 5-7

Hetero Diels-Alder reactions of 1-thia-1,3-butadienes have furthermore been employed for the construction of novel fullerene derivatives and in this work, Eguchi's group was successful in obtaining yields up to 69% of the novel fullerene adducts [432]. Thiochroman-fused fullerenes were also synthesised [433]; this method takes advantage of the in situ generation of the heterodiene – *o*-thioquinone – by thermal cycloreversion of benzothiet [434, 435]. Other current investigations concerning Diels-Alder reactions of 1-thia-1,3-butadienes deal with the synthesis of heterocycle-fused thiopyrans [436] and with exploring the reactivity of carbothionic esters as heterodienes [437].

In contrast to 1-thia-1,3-butadienes, the use of 2-thia-1,3-butadienes in hetero Diels-Alder reactions has only sparingly been studied. First work concerning the stereochemical course of intramolecular cycloadditions involving 2-thiabutadienes has been carried out by Beifuss et al. very recently [438]. The cationic 2-thia-1,3-butadiene **5-25** was formed upon treatment of the aldehyde **5-24** with thiophenol and underwent a cationic cycloaddition yielding three diastereomeric hexahydrothioxanthenes **5-26, 5-27** and **5-28** (Fig. 5-8).

Thus, 2-thia-1,3-butadienes have emerged as useful, nevertheless only little explored heterodienes which are expected to attract considerable interest in future research.

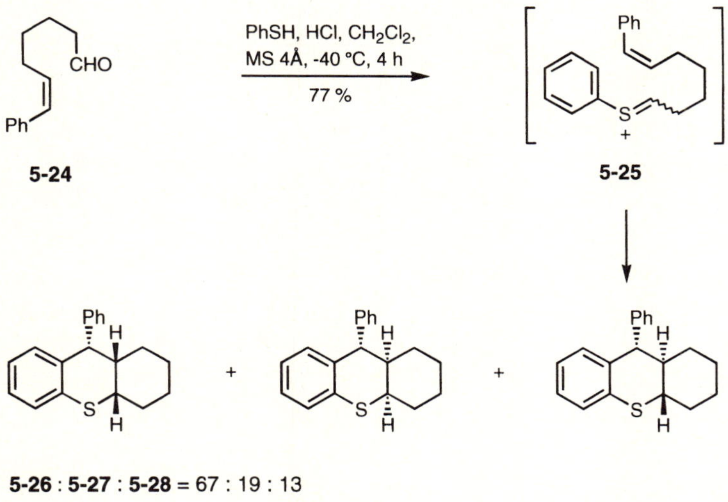

PhSH, HCl, CH_2Cl_2, MS 4Å, -40 °C, 4 h

77 %

5-24 **5-25**

5-26 : **5-27** : **5-28** = 67 : 19 : 13

Fig. 5-8

6
Miscellaneous Hetero Diels-Alder Reactions

In addition to the mentioned oxa-, aza- and thia-1,3-butadienes and the hitherto discussed dienophiles, a multitude of polyheterodienes and -dienophiles have been employed in recent hetero Diels-Alder reactions. The array of heteroatoms amenable for such cycloadditions is by no means restricted to oxygen, nitrogen and sulfur since there exist numerous reactions involving less common heteroatoms such as phosphorus, silicon or selenium.

A well-studied class of polyheterodienes is represented by 1-thia-3-aza-1,3-butadienes which yield 1,3-thiazine derivatives upon cycloaddition to suitable dienophiles [13]. They have been thoroughly investigated by Barluenga's group; these studies clearly demonstrate their high reactivity towards electron-deficient dienophiles [439] including azo compounds [440] and activated nitriles [441]. 1-Thia-3-aza-1,3-butadienes also undergo smooth intramolecular hetero Diels-Alder reactions with unactivated C=C double bonds [442]. Very recently, Guingant et al. has employed chiral dienophiles in order to open an asymmetric route to 1,3-thiazines [443]. Thus, the Lewis acid catalysed reaction of the heterodiene **6-1** with the Evans-type oxazolidinone derivative **6-2** yielded the cycloadduct **6-3** as a single diastereomer. Strikingly, under thermal or high pressure conditions a reversal of stereoselectivity occurred (Fig. 6-1).

Hetero Diels-Alder reactions of 1-thia-3-aza-1,3-butadienes have also attracted theoretical interest resulting in a study dealing with kinetic and thermodynamic parameters of such cycloadditions [444].

N-Acylimines which may react as 1-oxa-3-aza-1,3-butadienes represent a class of heterodienes which exhibit a close relationship to 1-thia-3-aza-1,3-butadienes [13]. A very impressive application of such an 1-oxa-3-aza-1,3-butadiene has been worked out by Swindell et al. [445]. The asymmetric hetero Diels-Alder reaction described therein opens a very elegant approach to the A-ring side chain of taxol. This synthesis takes advantage of the bulky chiral auxiliary attached to the dienophile **6-5** which upon cycloaddition with the 1-oxa-3-aza-1,3-butadiene **6-4** yielded the 1,3-oxazine derivative **6-6**. Subsequent hydrolysis, hydrogenolysis and transesterification gave the methyl ester of the taxol A-ring side chain **6-7** in good *endo* and excellent π-facial selectivity (Fig. 6-2).

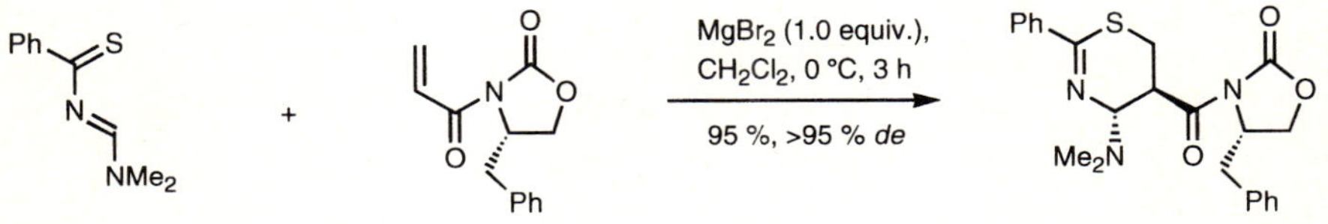

MgBr$_2$ (1.0 equiv.), CH$_2$Cl$_2$, 0 °C, 3 h
95 %, >95 % *de*

6-1 **6-2** **6-3**

Fig. 6-1

benzene, RT
75 %, 86 % *de*

1. H$_3$O$^+$
2. H$_2$/Pd(OH)$_2$
3. MeONa/MeOH

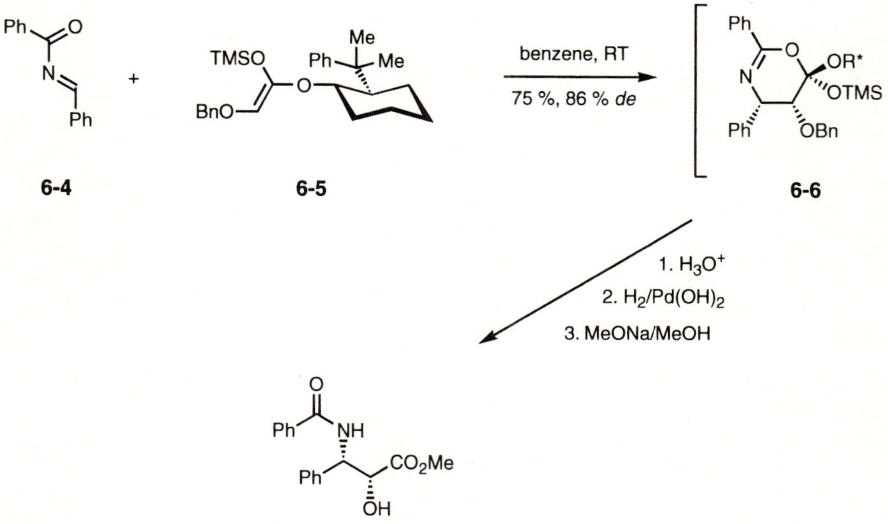

6-4 **6-5** **6-6** **6-7**

Fig. 6-2

Other, less extensively investigated bisheterodienes are 1,4-dioxa-, 1,4-dithia- and 1-oxa-4-thia-1,3-butadienes. Thus, several 1,4-benzodioxines have been synthesised by Nair et al. starting from *o*-benzoquinones [446, 447]. In an earlier work, Dondoni has studied their reactivity towards the C=C double bond of several oxazoles [448]. Cyclic dithiaoxalates have been employed as 1,4-dithia-1,3-butadienes in hetero Diels-Alder reactions yielding annellated 1,4-dithianes [449]. Very recently, also 1-oxa-4-thia-1,3-butadienes have been introduced as novel bisheterodienes [450]. They have been employed in a promising nonconventional glycosidation protocol presented by Franck et al. [451]. After being formed in situ from its precursor **6-8**, the heterodiene **6-9** cycloadded to the glycal **6-10**. Subsequent reductive cleavage of the resulting cycloadduct **6-12** yielded the glycoside **6-11** in moderate yield (Fig. 6-3).

Conceptionally related glycosidation methodologies stem from Leblanc's [452] and Schmidt's groups [453]. They involve azodicarboxylates (which actually are well known to react as heterodienophile, see Sect. 3.4.) as 1,2-diaza-4-oxa-1,3-butadienes; the resulting cycloadducts are valuable intermediates e.g. for the synthesis of *N*-acetyllactosamine [454].

Numerous studies have dealt with different types of sulfur-containing heterodienophiles. Thus, hetero Diels-Alder reactions of *N*-sulfinyl dienophiles have been thoroughly studied by Weinreb et al. [454]; the resulting cycloadducts represent useful and versatile intermediates in the synthesis of homoallylic amines [455] or pyrroles [456]. Further work using this type of S=N dienophiles

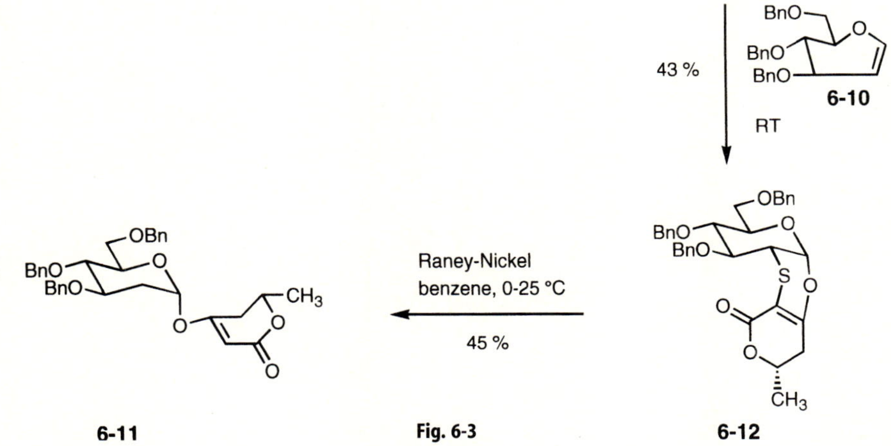

6-8 → pyridine → [6-9]; 6-9 + 6-10 (BnO, BnO, BnO), RT, 43 % → 6-12; 6-12 → Raney-Nickel, benzene, 0-25 °C, 45 % → 6-11

Fig. 6-3

has been directed towards the preparation of biotin [457]. The reactivity of such dienophiles strongly depends on an activation by an electron-withdrawing moiety attached to the nitrogen [454]; upon substitution with an electron-donating group, the application of high pressure or Lewis acid catalysis is required to induce a cycloaddition to a 1,3-diene [458]. It is noteworthy that a Lewis acid catalysed hetero Diels-Alder reaction of 1,4-dimethylbutadiene with the dienophile **6-14** yielded exclusively the cycloadduct **6-13**. However, under high pressure conditions **6-15** was formed as main product (Fig. 6-4).

Asymmetric hetero Diels-Alder reactions of *N*-sulfinyl dienophiles with chiral dienes have been found to proceed with very good induced diastereoselectivities [459].

N-Sulfonyl dienophiles are extremely reactive electrophiles which cannot be isolated. Nevertheless, a recent study carried out by Schaumann et al. reveals them to react with various carbo- and heterodienes in formal hetero Diels-Alder reactions [460].

Sulfur dioxide is known to readily undergo cheletropic reactions with 1,3-dienes for a long time. However, Vogel's group was successful in demonstrating that at low temperatures sulfur dioxide is indeed able to act as heterodienophile in hetero Diels-Alder reactions [461, 462]. Thus, isoprene and sulfur dioxide underwent a reversible cycloaddition to form sultine **6-16** as single regioisomer at – 60 °C. At – 40 °C, the well-known cheletropic reaction leading to the sulfolene **6-17** occurred (Fig. 6-5) [463].

Several ab initio studies of these reactions including an investigation of Lewis acid catalysis and solvent effects have been published by Sordo et al. [464–466]. Their results concerning regio- and stereoselectivity in the hetero Diels-Alder reactions of sulfur dioxide to isoprene are in good agreement with the experimental findings mentioned above.

Diatomic sulfur has also been shown to undergo hetero Diels-Alder reactions, and the complementary cycloreversion of suitable dithiines has been introduced as useful preparative method for the generation of this highly reactive form of sulfur [467, 468].

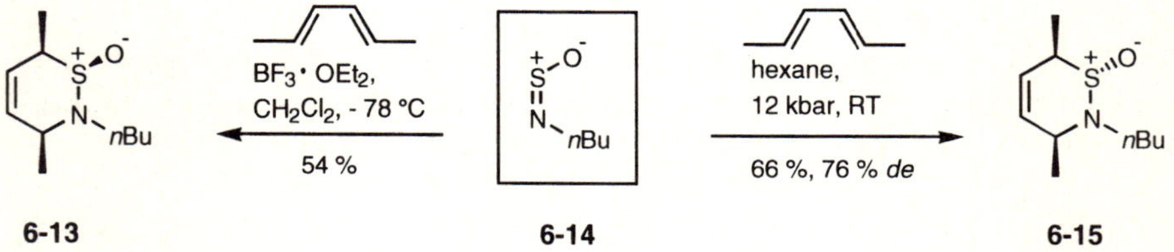

$BF_3 \cdot OEt_2$, CH_2Cl_2, - 78 °C, 54 %

hexane, 12 kbar, RT, 66 %, 76 % *de*

6-13 **6-14** **6-15**

Fig. 6-4

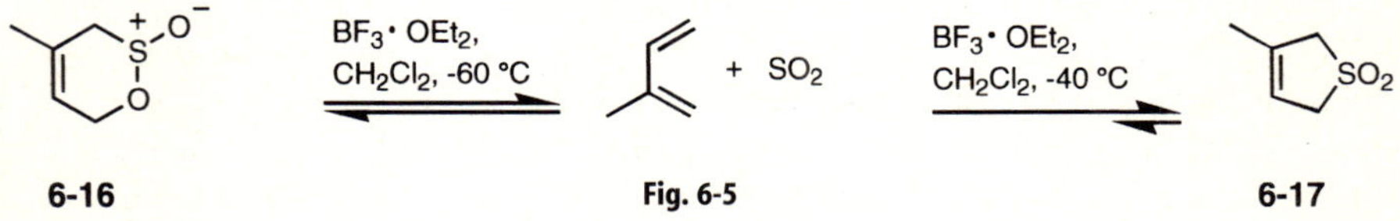

$BF_3 \cdot OEt_2$, CH_2Cl_2, -60 °C

+ SO_2

$BF_3 \cdot OEt_2$, CH_2Cl_2, -40 °C

6-16 **Fig. 6-5** **6-17**

Finally, some dienes and dienophiles bearing less common heteroatoms shall be discussed. Thus, Diels-Alder reactions involving phosphaalkynes as heterodienophilic component have extensively been used for the preparation of phosphabenzenes and Dewar-phosphabenzenes [469]. Another interesting approach to $3\lambda^3$, $3'\lambda^3$-diphosphinines (i.e. 3,3′-diphospha-biphenyles) and to other oligoaromatic, phosphorus-containing compounds has been developed by Märkl and his coworkers [470]. The cycloadditions described therein involve 1,3-azaphosphinines (which may be conceived as 1-phospha-3-aza-1,3-butadienes) as 4π component.

Some Sila-Diels-Alder reactions are also known; thus, the sterically stabilized silylidenephosphane **6-18** gave the adduct **6-19** in quantitative yield upon treatment with cyclopentadiene (Fig. 6-6) [471].

Sakurai et al. were successful in reacting even tetramethylsilene ($Me_2Si=SiMe_2$) with benzene in a photochemical [2+4] cycloaddition at 10 K in an argon matrix [472]. With mentioning the suitability of iminoboranes [473] and selenoaldehydes [474] to serve as dienophiles in hetero Diels-Alder reactions, this enumeration of exemplary, less usual hetero Diels-Alder reactions shall be completed.

$iPr_3Si{-}P{=}Si(iPr_3C_6H_2)_2$ → (toluene, RT, 5 h; 100 %) → $Si(iPr_3C_6H_2)_2$, P, $SiiPr_3$

6-18 **Fig. 6-6** **6-19**

7 Natural Product Syntheses by Hetero Diels-Alder Reactions

7.1 Syntheses with Oxa Diels-Alder Reactions

The dihydropyrans resulting from an oxa Diels-Alder reaction represent valuable intermediates for the synthesis of numerous natural compounds. In particular, they exhibit many structural elements of carbohydrates. It is therefore not surprising that both the normal electron demand cycloaddition of dienes to carbonyl dienophiles as well as the reaction of 1-oxa-1,3-butadienes with electron-rich alkenes have extensively been used for the synthesis of sugar derivatives. Nevertheless, various approaches to other natural products have been worked out by means of these powerful tools.

In early work which have been summarized very recently [23], Danishefsky et al. have investigated hetero Diels-Alder reactions of carbonyl compounds in order to yield glycals. Numerous further contributions to the stereoselective synthesis of dihydropyran derivatives by high pressure or Lewis acid induced Diels-Alder reactions of carbonyl compounds have been made by Jurczak et al., and as current application of this approach in natural product synthesis, his formal synthesis of compactin and mevinolin [89] has been discussed in Sect. 2.1.

This methodology which bases on Oppolzer's sultams has also been employed in the synthesis of 2,6-*N*,*N*-diacetyl-D-purpurosamidine C [475, 476]. Hetero Diels-Alder reactions of carbonyl dienophiles are furthermore involved in the preparation of (±)-ketoheptulosic acid [477] and in the already mentioned asymmetric approach to the C-26-C-32 tetrahydropyran subunit of swinholide A [92].

The inverse electron demand hetero Diels-Alder reaction of 1-oxa-1,3-butadienes and electron-rich dienophiles is an extremely versatile tool in natural product synthesis. This cycloaddition represents the key step of numerous approaches not only to carbohydrates, but also to terpenes, alkaloids, polyethers, steroid derivatives and various biologically active metabolites.

The advanced state of the art in carbohydrate synthesis basing on hetero Diels-Alder reactions of 1-oxa-1,3-butadienes has opened an access to enantiopure sugar derivatives. Thus, our group found the cycloaddition of the chiral heterodiene **7-1** and the electron-rich alkene **7-2** under the influence of Me_2AlCl to give the dihydropyran **7-3** in excellent *endo* selectivity (*endo*/*exo* >50:1) and as well excellent induced diastereoselectivity (54:1) [478]. A short sequence involving one simple recrystallisation then led to the ethyl-β-D-mannopyranoside **7-4** in enantiomerically pure form (Fig. 7-1).

Two aspects of this work are noteworthy: Firstly, the excellent induced diastereoselectivity results from a very remote inducing stereocenter (1,6-asymmetric induction) and secondly, the asymmetric induction can be reversed by changing the Lewis acid. Thus, the sequence also allows the entrance into the unnatural L-series by performing the cycloaddition with the same substrates but in the presence of TMSOTf.

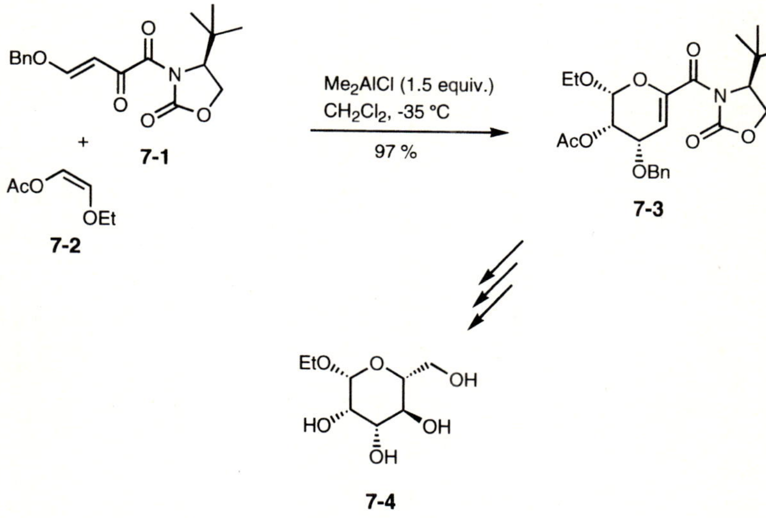

Fig. 7-1

The use of chiral 1-oxa-1,3-butadienes for the stereoselective preparation of carbohydrates has also been investigated by Schmidt et al. [479]. The elegant syntheses of *N*-acetyl-β-D-neuraminic acid derivatives are an impressive result of these studies [480].

A very recent study presented by Dujardin et al. describes the complementary use of chiral enol ethers as dienophiles in oxa Diels-Alder reactions. This approach has yielded promising results with regard to the synthesis of enantiomerically pure carbohydrates [481]. Further noteworthy studies directed to the preparation of biologically active amino sugars from enaminoketones have been carried out in our group [110].

There are numerous indole alkaloids known which bear a dihydropyran moiety. Amongst them, strictosidine plays an outstanding role as biosynthetic intermediate more than two thousand natural alkaloids are derived from. Our group has synthesised analoga of this important natural product by the highly efficient domino-Knoevenagel-hetero Diels-Alder protocol (see Sect. 2.2) and was successful in converting the resulting cycloadducts into dihydrocorynantheine derivatives [482].

(–)-Tetrahydroalstonine **7-7**, a heteroyohimboid alkaloid, has been synthesised in enantiopure form by Martin et al. by means of an oxa Diels-Alder reaction as key step. The trienic precursor **7-5** underwent a thermal intramolecular cycloaddition to form a 5:1 mixture of **7-6** and its 15β-epimer. The main cycloadduct was then subjected to a straightforward sequence to yield the natural product **7-7** (Fig. 7-2) [483–485]. In earlier work, Ogasawara et al. have employed a conceptionally different domino Knoevenagel-hetero Diels-Alder approach to this alkaloid and other natural products [486–488].

Our powerful domino-Knoevenagel-hetero Diels-Alder procedure has also proven its value in the synthesis of terpene derivatives. Thus, 1-ethoxysecologa-

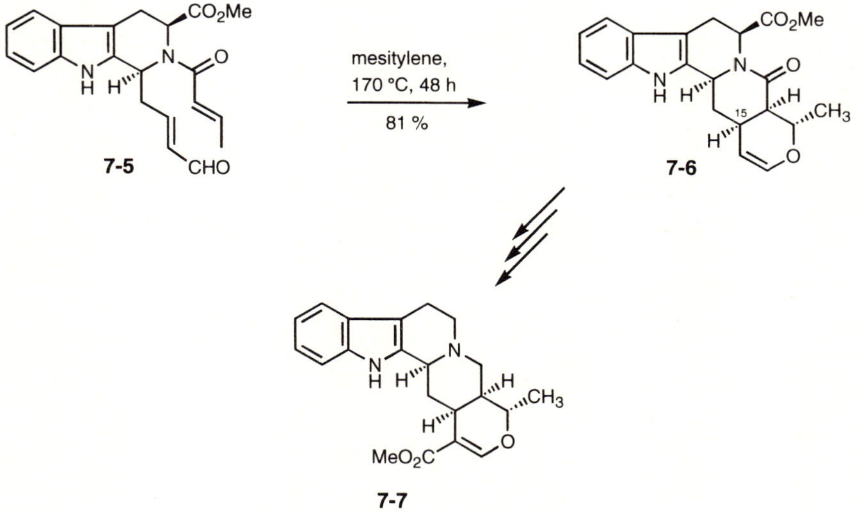

Fig. 7-2

nin-aglycon **7-12** has been prepared by means of an extremely efficient three component reaction which involved **7-8** and **7-9** as substrates for the formation of the intermediate 1-oxa-1,3-butadiene by Knoevenagel condensation and **7-10** as dienophile to be employed in the subsequent cycloaddition. The cycloadduct **7-11** was then easily converted into the secologanin derivative **7-12** which was isolated in 12% overall yield [489] (Fig. 7-3).

Secologanin **7-13** is a direct precursor of strictosidine and is therefore of outstanding importance with regard to the biosynthesis of alkaloids. Furthermore, its formyl moiety itself could also been subjected to the domino-Knoevenagel-hetero Diels-Alder protocol to yield bridged homoiridoids [490]. Another important application of this methodology to the chemistry of terpenes is the synthesis of deoxyloganin [491].

The domino-Knoevenagel-hetero Diels-Alder reaction is furthermore suitable for the efficient preparation of D-homosteroids [492]. Another effective use of this method is the synthesis of heterosteroids, which are interesting due to their potential pharmacological properties [493].

Spiroketals are not only important building blocks of polyethers but also may represent themselves highly active natural products. The suitability of oxa Diels-Alder reactions to efficiently generate this structure will be demonstrated by two impressive examples. Thus, our group prepared the mycotoxine (–)-talaromycin B **7-17** by a nine-step synthesis in 5% overall yield in enantiopure form. The

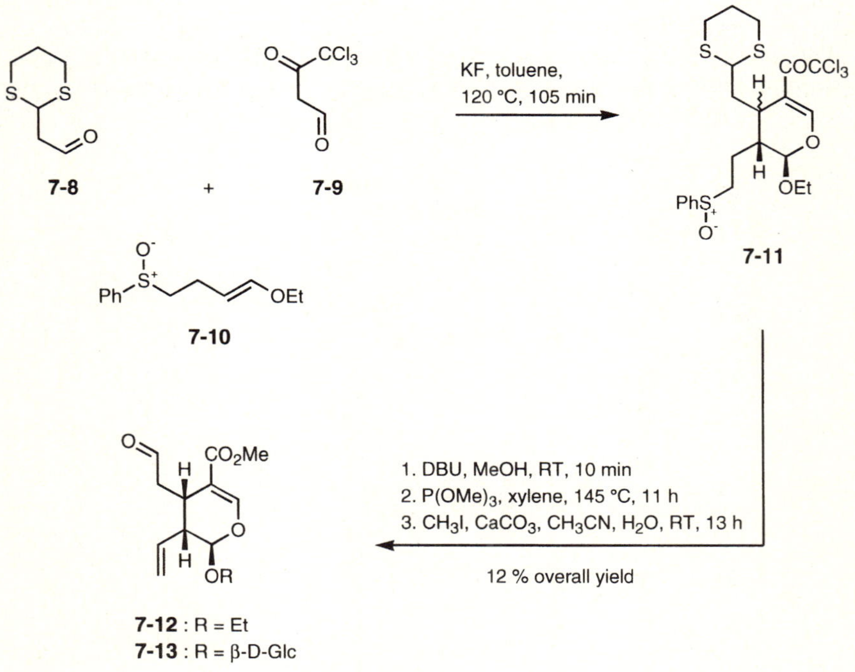

KF, toluene, 120 °C, 105 min

1. DBU, MeOH, RT, 10 min
2. $P(OMe)_3$, xylene, 145 °C, 11 h
3. CH_3I, $CaCO_3$, CH_3CN, H_2O, RT, 13 h

12 % overall yield

7-12 : R = Et
7-13 : R = β-D-Glc

Fig. 7-3

chiral enol ether **7-14** underwent an oxa Diels-Alder reaction with the heterodiene **7-15** to give **7-16** as main product amongst four diastereomeric cycloadducts. Although complete separation of **7-16** from its byproducts was not possible at this stage, four subsequent steps yielded chemically and optically pure (–)-talaromycin B (Fig. 7-4) [494].

Recent work from Ireland's group directed to the synthesis of monensin subunits takes as well advantage of the suitability of hetero Diels-Alder reactions for generating spiroketals [495]. In situ generation of the highly labile dienophile **7-19** from **7-18**, subsequent cycloaddition to acrolein which acted as 1-oxa-1,3-butadiene and reduction yielded the spiroketal **7-21** accompanied by small amounts of diastereomeric byproducts. A mild acid catalysed rearrangement is the next key transformation to the spiroketal subunit **7-20** of monensin **7-22** (Fig. 7-5).

Snider et al. have synthesised the antiinsectan (±)-leporin [496] **7-26** using the domino-Knoevenagel-hetero Diels-Alder sequence. The intermediate 1-oxa-1,3-butadiene **7-25** was formed in situ by condensation of the pyridone **7-23** and the dienal **7-24**. Subsequently, a hetero Diels-Alder reaction occurred accompanied by minor side reactions. Thus, the desired cycloadduct **7-27** was formed only in moderate yield as 5:1 mixture with its *trans*-fused diastereomer (Fig. 7-6). Functionalisation of the nitrogen atom yielded the natural product. A similar reaction sequence occurred in the synthesis of the structurally related free radical scavenger (±)-pyridoxatin, however, in this approach the hetero Diels-Alder reaction represented only a side reaction competing with the desired intramolecular ene reaction [497].

Beyond the hitherto discussed syntheses, there exist numerous further current and impressive applications of Diels-Alder reactions involving 1-oxa-1,3-butadienes which can only briefly be mentioned here due to the limits of space. Burke's group has developed an elegant retro-oxa Diels-Alder/all-carbon-Diels-Alder protocol which was employed as key step in the total syntheses of (±)-

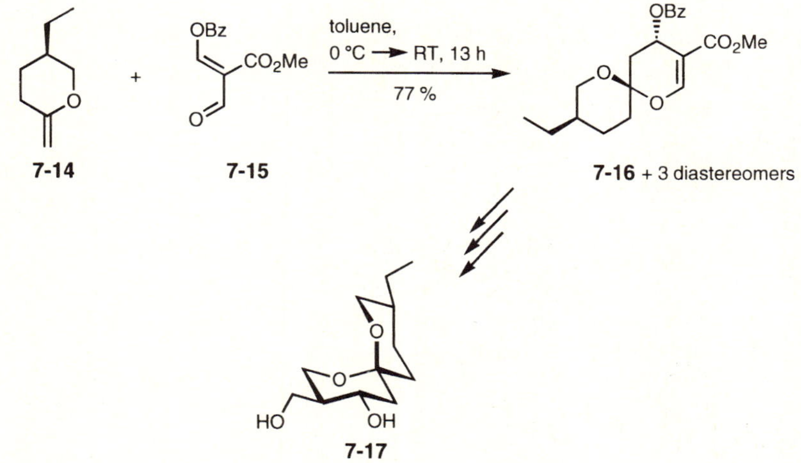

Fig. 7-4

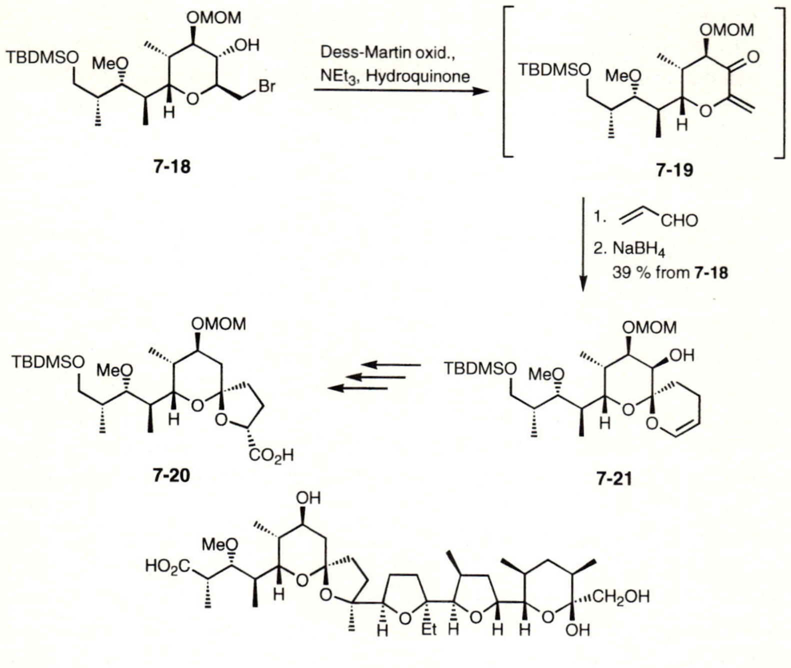

TBDMSO MeO OMOM OH H O Br

7-18

Dess-Martin oxid., NEt_3, Hydroquinone

TBDMSO MeO OMOM O H O

7-19

1. CHO
2. $NaBH_4$
39 % from **7-18**

TBDMSO MeO OMOM H O O CO_2H

7-20

TBDMSO MeO OMOM OH H O O

7-21

OH MeO HO_2C H O O H O Et H O H H O CH_2OH OH

7-22

Fig. 7-5

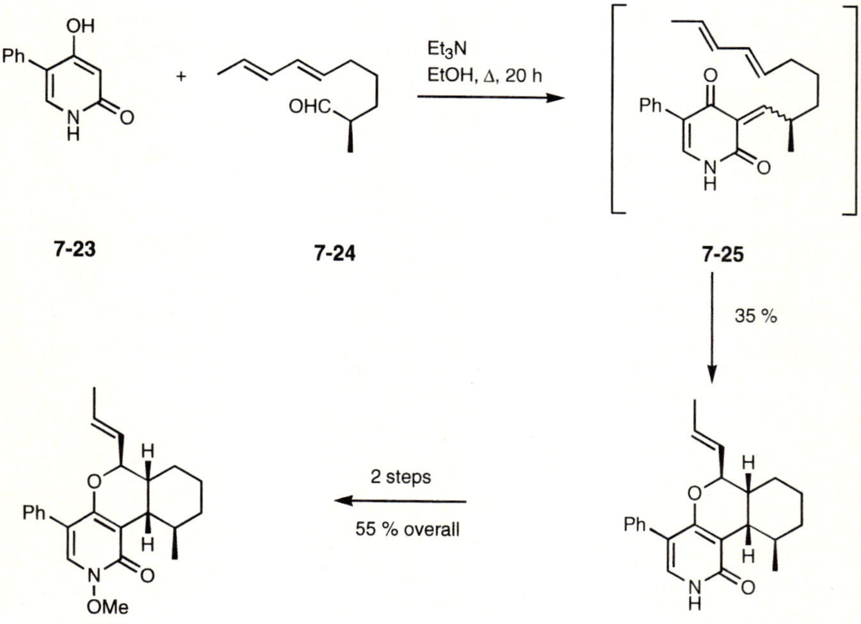

Ph OH N H O

7-23

\+

OHC

7-24

Et_3N
EtOH, Δ, 20 h

O Ph N H O

7-25

35 %

Ph O H H N H O

7-27

2 steps
55 % overall

Ph O H H N OMe O

7-26

Fig. 7-6

pulo'upone [498] and of the ionophore antibiotic indanomycin [499, 500]. Other applications of 1-oxa-1,3-butadienes have been aimed at the preparation of cannabinoids [501, 502], carbapanems [503] and the antibiotic ramulosin [504].

7.2 Syntheses with Aza Diels-Alder Reactions

Hetero Diels-Alder reactions with imino dienophiles have been employed as key step in several syntheses of naturally occuring alkaloids. With regard to stereoselective transformations, the approach to (*S*)-anabasin worked out by Kunz et al. impressively illustrates the high utility of natural carbohydrates as source of chirality in asymmetric synthesis [505]. The *N*-galactosyl imine **7-28** underwent a Lewis acid catalysed aza Diels-Alder reaction with Danishefsky's diene which proceeded with excellent induced diastereoselectivity to yield the adduct **7-29**. A short sequence then afforded the desired alkaloid **7-30**. This work also deals with the suitability of several other dienes and imino dienophiles for such transformations (Fig. 7-7).

Two very elegant alkaloid syntheses basing on intramolecular cycloadditions of imino dienophiles have been published by Grieco and his coworkers. The preparation of (±)-eburnamonine **7-32** is very efficient since imine **7-31**, available from δ-valerolactam in a straightforward sequence, is directly converted into the desired alkaloid **7-32** by aza Diels-Alder reaction and subsequent isomerisation of the newly formed double bond. (Fig. 7-8) [506].

The in situ release of a reactive iminium ion by cycloreversion of an azanorbornene [211, 212] and a subsequent intramolecular aza Diels-Alder reaction

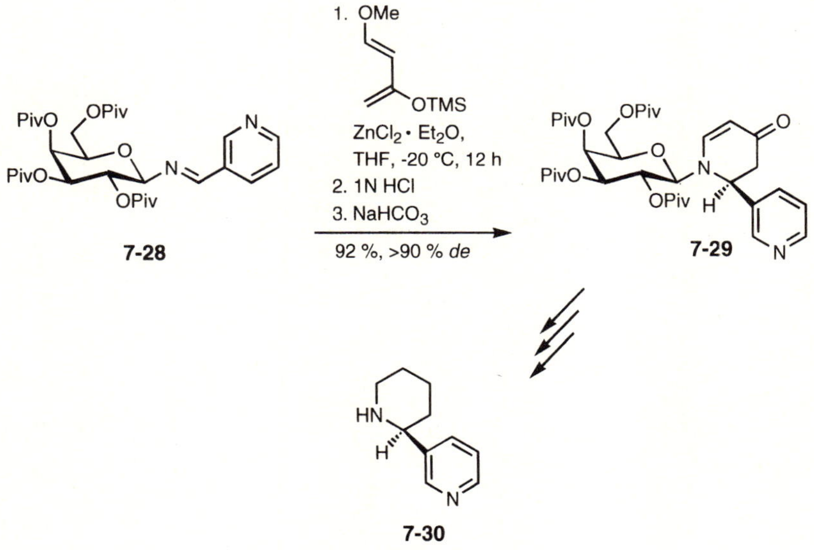

Fig. 7-7

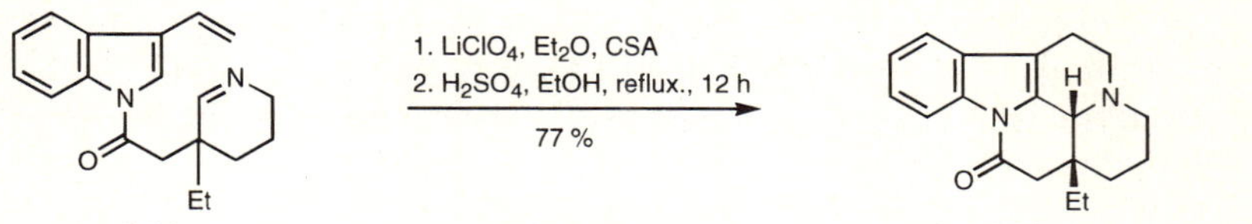

1. $LiClO_4$, Et_2O, CSA
2. H_2SO_4, EtOH, reflux., 12 h
77 %

7-31 **Fig. 7-8** 7-32

represent one important key step of the biomimetic total synthesis of (±)-pseudotabersonine **7-37** [507]. Thus, upon heating **7-33** in the presence of a Lewis acid, formation of the two diastereomeric cycloadducts **7-35** and **7-36** with the intermediacy of the iminium ion **7-34** occurred. The main product **7-35** was then transformed to the alkaloid **7-37**, and it is noteworthy that the remaining fifth ring was formed by an intramolecular all-carbon Diels-Alder reaction which completely established the relative configuration of pseudotabersonine. Since all present stereogenic centers had been destroyed before this cycloaddition, the low stereoselectivity of the previous aza Diels-Alder reaction proved to be no serious drawback (Fig. 7-9).

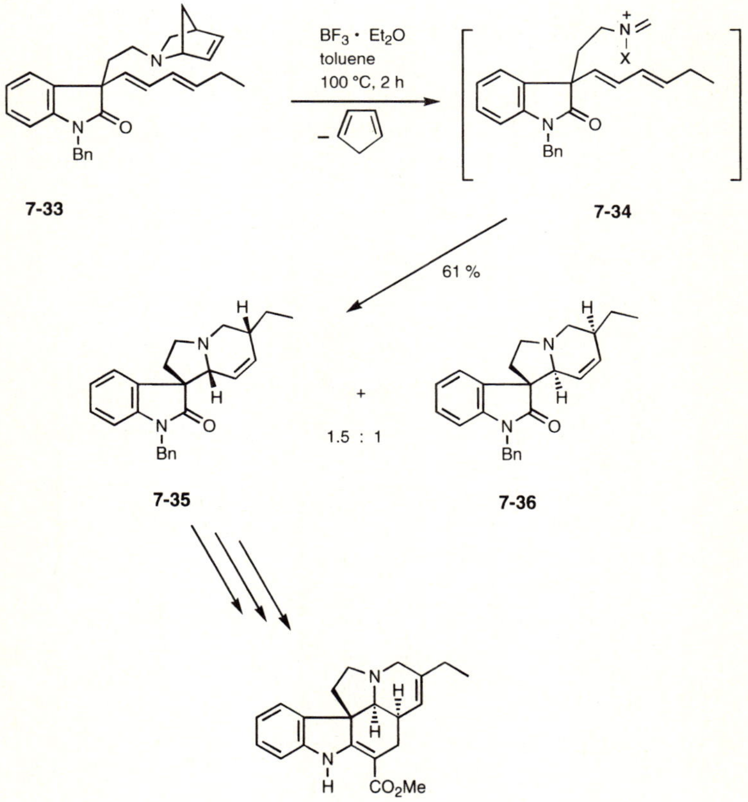

$BF_3 \cdot Et_2O$
toluene
100 °C, 2 h

7-33 7-34

61 %

1.5 : 1

7-35 7-36

Fig. 7-9

Further applications of imino dienophiles to the synthesis of natural or biologically active compounds have been directed to (–)-cannabisativine [508] and HIV-1 protease inhibitors [509]; Bailey's investigations of the enantioselective synthesis of pipecolic acid derivatives have already been discussed in Sect. 3.1.

Syntheses of natural products involving aza Diels-Alder reactions may also employ aza-1,3-butadienes as nitrogen containing component. Amongst the 1-aza-1,3-butadienes, Boger's *N*-sulfonyl-1-azabuta-1,3-dienes have recently proven to be versatile intermediates for antitumor agents such as streptonigrone [510] and fredericamycin A [511]. Whilst in the latter total synthesis the hetero Diels-Alder reaction represents a very early key transformation as part of a convergent synthetic strategy, the cycloaddition of the doubly activated heterodiene **7-38** with the ketene acetal **7-39** effects the completion of the tetracyclic streptonigrone skeleton. The resulting cycloadduct was directly subjected to an aromatization protocol leading to **7-40**; a short sequence then afforded the natural product **7-41** (Fig. 7-10).

The preparation of amphimedine **7-46** published by Echavarren and Stille [512] is a noteworthy application of 2-aza-1,3-butadienes in natural product synthesis since it is an interesting combination of hetero Diels-Alder methodology with a palladium catalysed cross coupling. Thus, dienophile **7-44** was formed by Stille coupling of the triflate **7-42** with the stannyl aniline **7-43**. This quinone then underwent cycloaddition to the 2-aza-1,3-butadiene **7-45**; an acid catalysed hydrolysis of the cycloadduct **7-47** and subsequent *N*-methylation completed the synthesis of amphimedine **7-46** (Fig. 7-11).

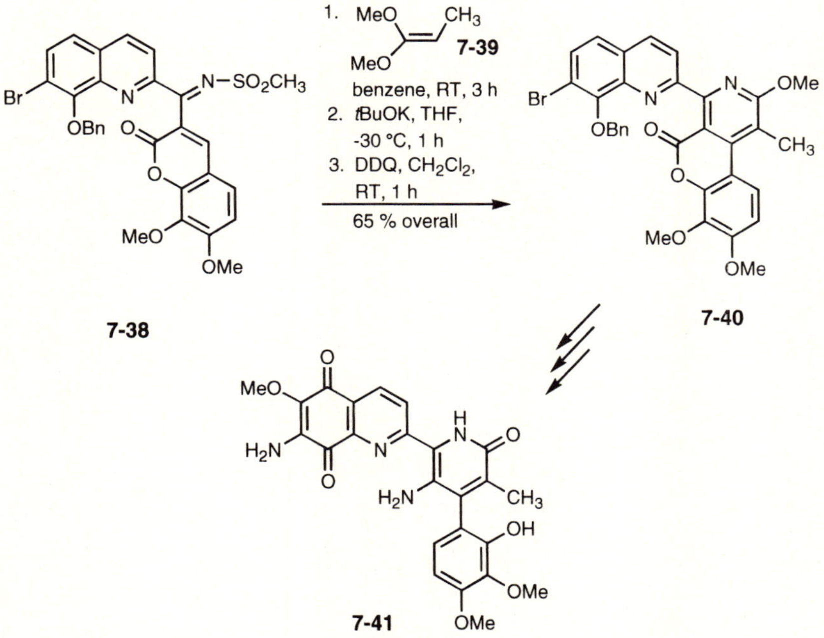

Fig. 7-10

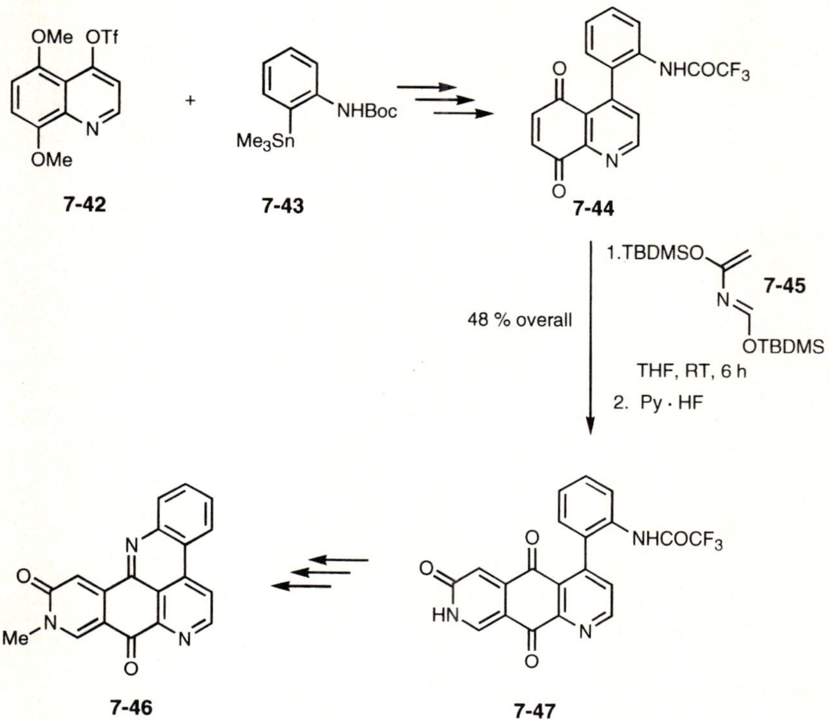

Fig. 7-11

Another hetero Diels-Alder reaction of a 2-aza-1,3-butadiene is part of Heathcock's extremely efficient polycyclization protocol leading to the skeleton of the Daphnyllum alkaloids [513]. This powerful sequential transformation is started by the generation of the highly reactive dialdehyde **7-49** from the corresponding diol **7-48.** Upon treatment with ammonia and subsequent protonation, the cationic 2-aza-1,3-butadiene **7-50** is formed which then undergoes an intramolecular cycloaddition to yield the iminium ion **7-51.** This intermediate is converted to the polycyclic final product **7-52** by an aza ene reaction. It is noteworthy that a closely related polycylisation cascade leading to the saturated analogon of **7-52** occurs if ammonia is replaced by methylamine. Heathcock et al. have applied these unique synthetical tools to the total syntheses of (+)-codaphniphylline **7-53** [514], (+)-daphnilactone A [515], bukittinggine [516], (–)-secodaphniphylline [517] and a number of related products (Fig. 7-12).

Franck's preparation of (–)-cryptosporin is an interesting natural product synthesis involving an isoquinolinium salt as cationic 2-aza-1,3-butadiene [518].

Boger et al. have worked out numerous total syntheses of natural products basing on aza Diels-Alder reactions of electron-deficient *N*-heterocycles which act as diaza-1,3-butadienes [11]. The key steps of these reactions have been highlighted in Sect. 3.5.

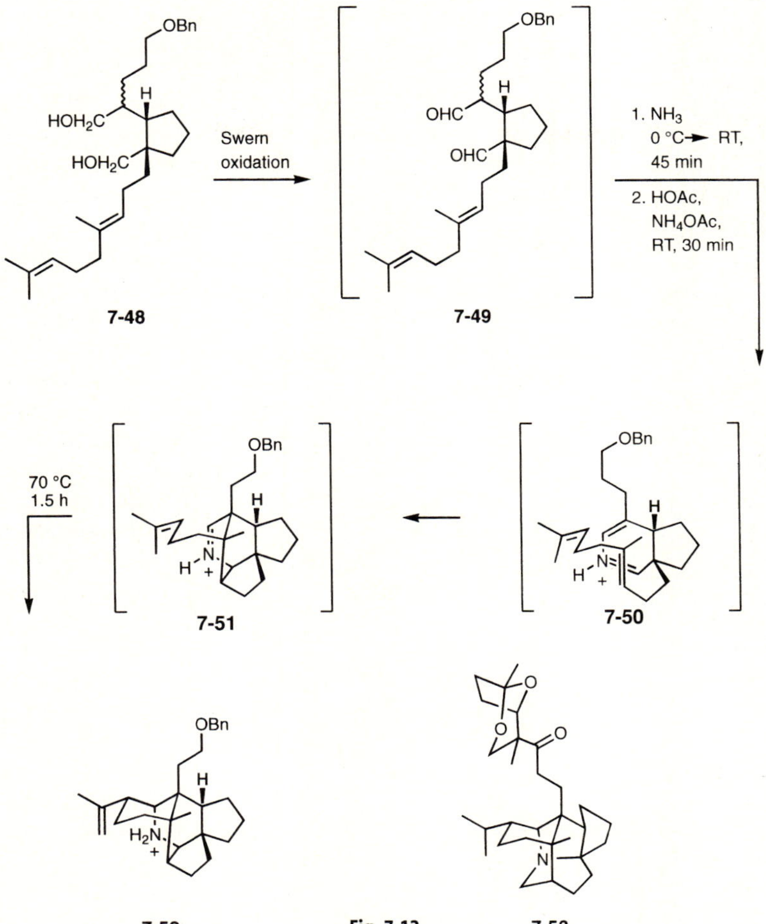

Fig. 7-12

7.3
Syntheses with Miscellanous Hetero Diels-Alder Reactions

The outstanding versatility of 1,2-oxazine derivatives resulting from hetero Diels-Alder reactions of nitroso dienophiles has been exploited in a multitude of natural product syntheses which have been reviewed recently [8]. Therefore the discussion in this paper shall focus on some very recent, typical applications of nitroso dienophiles.

An asymmetric Diels-Alder reaction of a chiral nitroso dienophile has been employed by Ganem et al. in order to open an elegant access to enantiomerically pure (+)-mannostatin A **7-57** and several derivatives thereof [519]. The cycloaddition of the heterodienophile **7-54** derived from mandelic acid to 1-methylthiocyclopentadiene **7-55** proceeded only in moderate diastereoselectivity, however, the desired product **7-56** was easily separated from its diastereomer.

It is noteworthy that both components are highly reactive and both required an in situ generation. The cycloadduct **7-56** was then subjected to a reductive ring contraction, acetylation, and a highly diastereoselective bishydroxylation to yield **7-58** which upon several protecting group transformations gave mannostatin A **7-57** in enantiomerically pure form (Fig. 7-13).

Streith and Defoin have employed nitroso dienophiles for numerous preparations of azasugars [392]. A very recent publication of this series describes the synthesis of glycosyl transferase inhibitors such as 1,6-dideoxynojirimycin **7-65**[520, 521]. This approach as well takes advantage of a chiral heterodienophile, namely the mannose derivative **7-60** developed by Kresze and Vasella. Upon cycloaddition of **7-60** and **7-59**, a 85:15 mixture of *cis*- and *trans*-**7-61** was formed which was directly *N*-protected to give **7-62**. The chiral auxiliary was easily recovered as mannolactone **7-63** after the hetero Diels-Alder reaction. The mixture of diastereomers **7-62** was chemically resolved since only *cis*-**7-62** reacted in the subsequent bishydroxylation step to give enantiopure **7-64** (however, as *E*/*Z*-mixture of the oximes). Configurational inversion via a cyclic sulfate which then was cleaved by nucleophilic attack yielded **7-66** as main product. Conversion into 1,6-dideoxynojirimycin **7-65** was accomplished by reductive *N*-deprotection and hydrogenolysis (Fig. 7-14).

Many further important natural product syntheses are covered by the aforementioned review. Particularly noteworthy amongst them are Hudlicky's syntheses of conduramines [522] and (+)-lycoricidine [523] since they employ enantioselective microbial oxidations of halobenzenes as source of chirality. Racemic lycoricidine has also been prepared by Martin et al.; this synthesis exhibits an interesting Heck cyclisation as key step in addition to the hetero Diels-

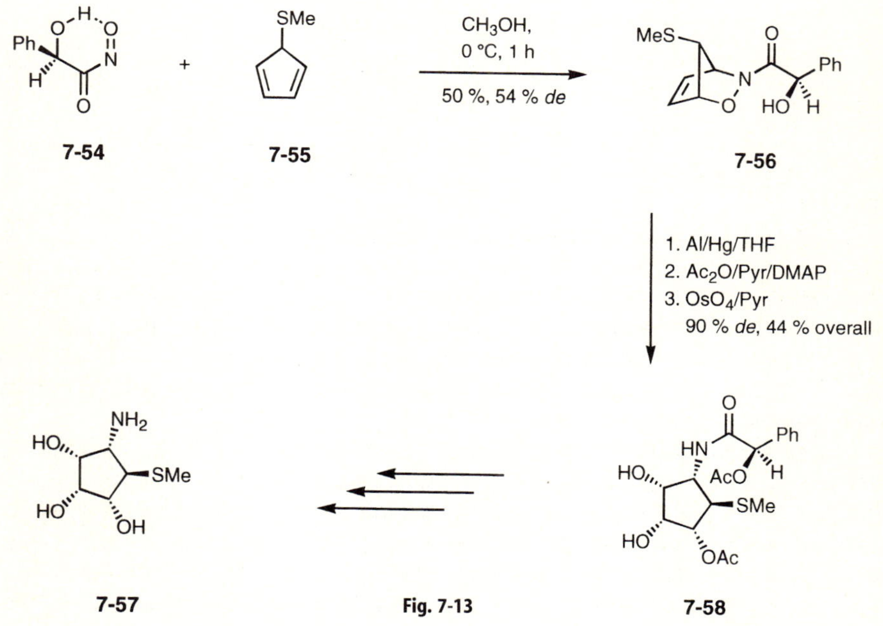

Fig. 7-13

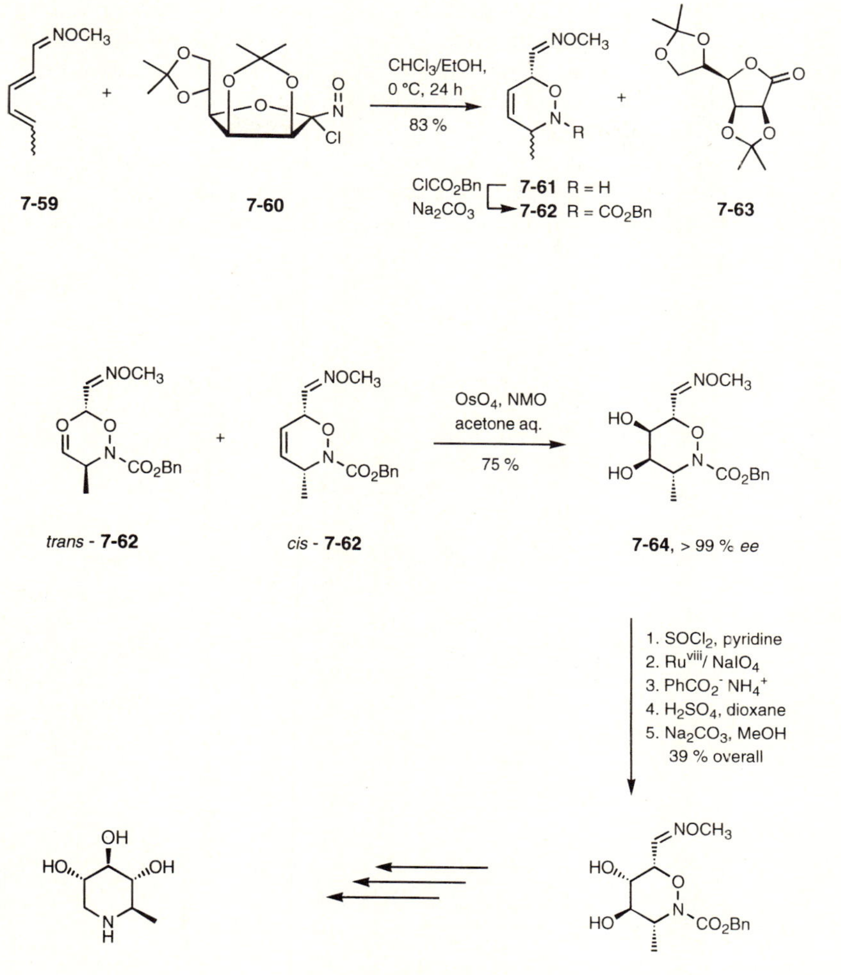

$CHCl_3$/EtOH, 0 °C, 24 h

83 %

$ClCO_2Bn$, Na_2CO_3

7-59 7-60 7-61 R = H 7-62 R = CO_2Bn 7-63

OsO_4, NMO, acetone aq.

75 %

trans - **7-62** *cis* - **7-62** **7-64**, > 99 % *ee*

1. $SOCl_2$, pyridine
2. Ru^{viii}/ $NaIO_4$
3. $PhCO_2^-$ NH_4^+
4. H_2SO_4, dioxane
5. Na_2CO_3, MeOH

39 % overall

7-65 **7-66**

Fig. 7-14

Alder reaction [524, 525]. Danishefsky et al. have used nitroso dienophiles for the synthesis of mitomycin K and antibiotics of the FR 900482 family, the latter ones are structurally unique aziridino-1,2-oxazine derivatives [526–529]. An approach directed to the cephalotaxus alkaloids has been worked out by Fuchs et al. [530], and several indolizidine alkaloids have been prepared by Keck's [531] and Kibayashi's groups [532, 533]. Kibayashi et al. also synthesised Nuphar piperidine alkaloids in enantiomerically pure form by means of an asymmetric nitroso Diels-Alder reaction [534].

Cycloadditions involving nitroalkenes as heterodiene have been employed as part of Denmark's domino [4 + 2] / [3 + 2] cycloaddition protocol for the synthesis of natural products. Since also this methodology has just been exhaustively reviewed [5], its value for alkaloid synthesis might be exemplarily demonstrated by discussing the asymmetric synthesis approach to (–)-rosemarinecine **7-71** [535]. Starting from the nitroalkene **7-67** and the chiral enol ether **7-68**, the [4+2]/[3+2] process catalysed by the bulky Lewis acid methylaluminium-bis(2,6-diphenylphenoxide) installed all stereogenic centers present in the natural product. The transformation of cycloadduct **7-69** into the natural product involved a reductive ring contraction to construct the pyrrolizidine skeleton **7-72** (which allowed as well essentially complete recovery of the chiral auxiliary **7-70**) and the cleavage of the lactol by means of Red-Al as key steps (Fig. 7-15).

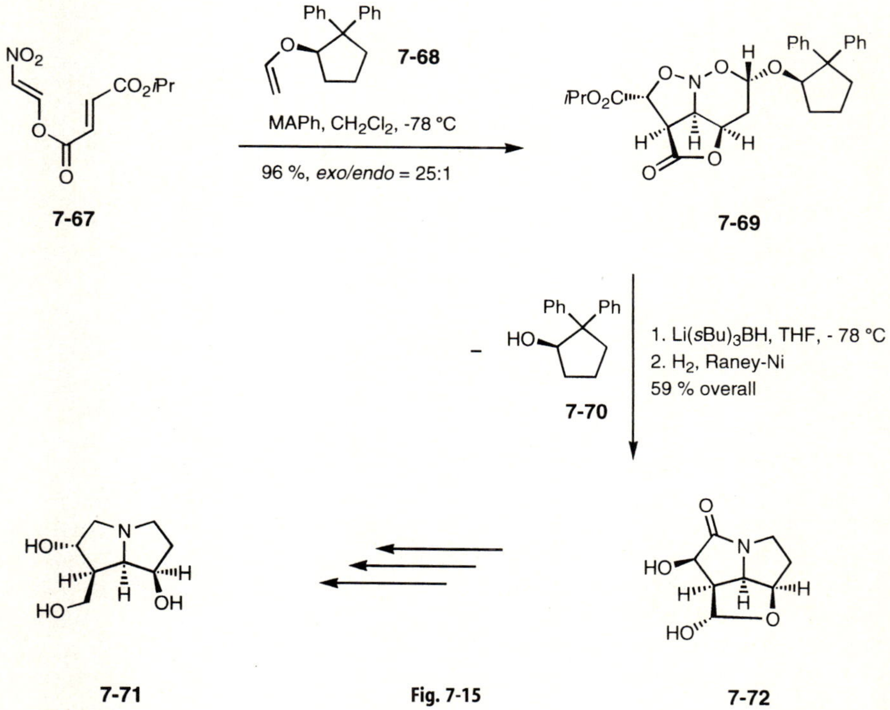

Fig. 7-15

Finally, a thia Diels-Alder reaction representing a less common cycloaddition type in natural product synthesis shall be discussed. Thus, Vedejs et al. have included such a cycloaddition into an elegant strategy aimed at the synthesis of macrocyclic [11]-cytochalasans such as zygosporin E **7-76** [536–538]. Thus, release of the thioaldehyde **7-73** from its phenacyl sulfide precursor in the presence of the silyloxydiene **7-74** yielded **7-75** as 2:1 mixture with its C_{20} epimer. Fortunately, equilibration of this mixture raised the ratio up to 10:1. Several subsequent steps yielded the tetracyclic intermediate **7-77**; cleavage of its thioether moiety then liberated the 11-membered macrocycle present e.g. in zygosporin *E* **7-76** (Fig. 7-16).

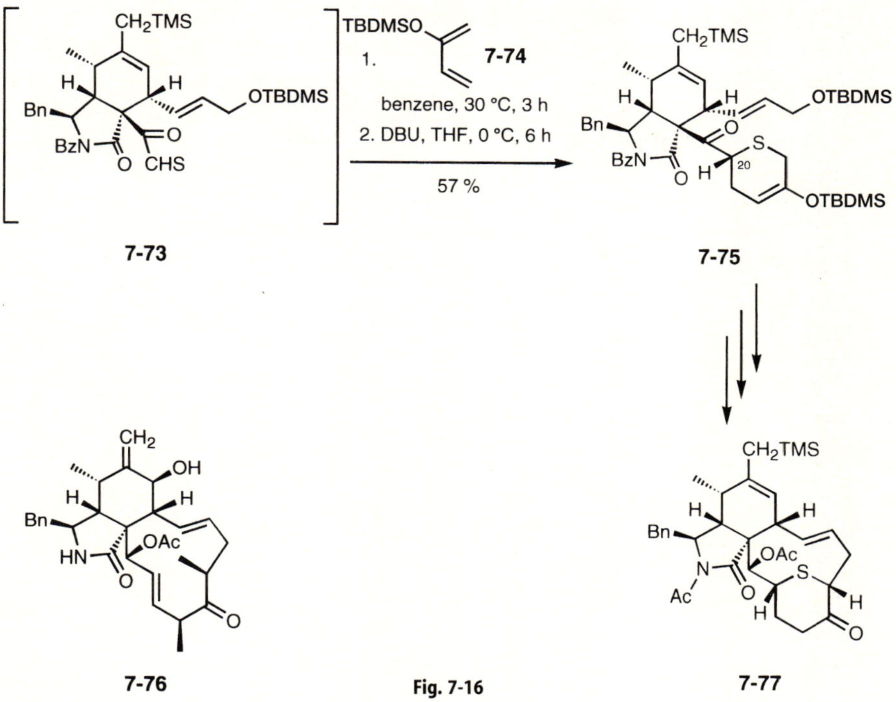

1. 7-74, benzene, 30 °C, 3 h
2. DBU, THF, 0 °C, 6 h
57 %

7-73 7-75

7-76 7-77

Fig. 7-16

8 High Pressure Applications in Hetero Diels-Alder Reactions

High pressure has developed as an efficient tool in organic synthesis especially in cases where the substrates are sensitive and large negative $\Delta V^{\#}$ can be expected as in cycloaddition reactions. The older work on high pressure application in hetero Diels-Alder reactions focused mainly on the increase of the reaction rate, whereas in newer work in addition the influence of high pressure on the selectivity is studied. Moreover, careful measurements of the kinetics and the determination of $\Delta V^{\#}$, $\Delta S^{\#}$, $\Delta H^{\#}$, and $\Delta\Delta V^{\#}$ have been performed for cycloadditions of 1-oxa-1,3-butadienes to allow also a mechanistic interpretation. As already mentioned, an increase of the reaction rate can also be obtained by using Lewis acids in many cases, thus, this method shows a similar effect to the application of high pressure [539]. However, the mechanistic reasons are completely different and there are several examples where only the application of high pressure allowed a transformation [e.g. 64].

Although several research groups have investigated the effect of high pressure on all-carbon Diels-Alder reactions, there are much less examples for the hetero Diels-Alder reaction. Thus, mainly only two groups are working on high pressure hetero Diels-Alder reactions. Thus, Jurczak has worked on the cycloaddition of oxa-dienophiles, whereas we are investigating the reaction of 1-oxa-1,3-butadienes under high pressure in co-operation with Buback.

Recently, Jurczak et al. analysed the Eu(fod)$_3$-mediated high pressure cycloaddition of 1-methoxy-1,3-butadienes **8-1** to chiral aminoaldehydes **8-2** with different protecting groups. There was no increase in selectivity by applying high pressure. Thus, reaction of **8-1 a** and **8-2** either in the presence of Eu(fod)$_3$ or without the Lewis acid gave **8-3 a** and **8-4 a** in the same ratio. However, the change of the protecting groups at the nitrogen led from a chelate to a non-chelate control to give **8-3** as the major product (Fig. 8-1) [540]. Similar effects on changing the protecting groups at the nitrogen had already been discussed earlier (see Sect. 2.1).

Comparable results were also obtained in the hetero Diels-Alder reaction of **8-1** and **8-5**. Again, the ratio of the products **8-6** and **8-7** was not changed under high pressure (Fig. 8-2) [541].

In a similar fashion, Achmatowicz et al. have studied the cycloaddition of 1-acetoxy- and 1-trimethylsilyloxy-3-methyl-1,3-butadiene with diethyl oxomalonate and isopropyl glyoxylate under thermal and high pressure conditions [542].

Vandenput et al. applied high pressure to perform hetero Diels-Alder reactions with unactivated 1-oxa-1,3-butadienes such as **8-8** and enol ethers **8-9** in the presence of a weak Lewis acid. The cycloadditions led to the dihydropyrans **8-10** as a mixture of diastereomers in 23 to 85% yield (Fig. 8-3) [543].

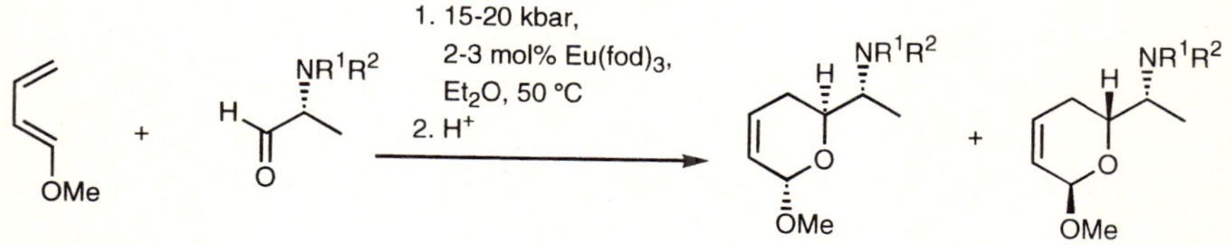

1. 15-20 kbar, 2-3 mol% Eu(fod)$_3$, Et$_2$O, 50 °C
2. H$^+$

8-1 + **8-2** → **8-3** + **8-4**

a: R^1R^2 = H,Cbz
b: R^1R^2 = Pht
c: R^1R^2 = Bn,Boc

	8-3		8-4
a:	1	:	2.5
b:	5	:	1
c:	16	:	1

Fig. 8-1

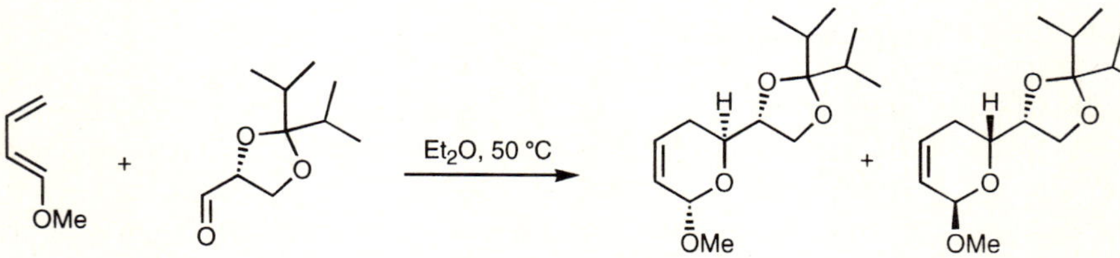

8-1 + **8-5** → (Et$_2$O, 50 °C) **8-6** + **8-7**

20 kbar + Eu(fod)$_3$ = 86 : 14 (64 %)
20 kbar = 84 : 16 (57 %)

Fig. 8-2

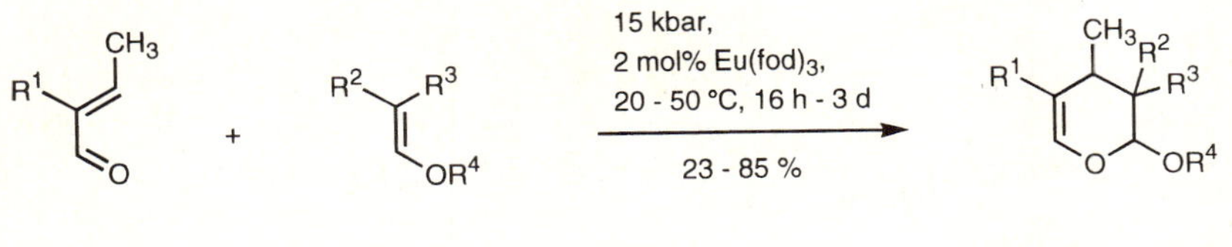

15 kbar, 2 mol% Eu(fod)$_3$, 20 - 50 °C, 16 h - 3 d; 23 - 85 %

8-8 + **8-9** → **8-10**

R^1 = H, CH_3 ; R^2 = CH_3 ; R^3 = H, CH_3 ; R^4 = Et, *i*Pr ; R^3R^4 = $(CH_2)_3$

Fig. 8-3

As already discussed in Sect. 2.2, the reactivity of 1-oxa-1,3-butadienes can be enhanced by introducing an electron-withdrawing group at the 2-position. But, with an electron-donating group such as a methoxy group at the 4-position the reactivity drops again greatly. However, the cycloadditions of these compounds can be accelerated by applying high pressure, besides using Lewis acids. Thus, reaction of **8-11** and **8-12** at 24 °C and 13 kbar led to the diastereomeric cycloadduct **8-13** in 82% yield as shown by Boger et al. (Fig. 8-4) [544]. A change of the diastereoselectivity under high pressure was not detected. This is consistent with earlier attempts to increase diastereoselectivity by applying high pressure; but the observed differences in activation volumes did not exceed 1 $cm^3\ mol^{-1}$.

As the first example of a pressure induced change in selectivity we have found a significant increase in diastereoselectivity by applying high pressure for the cycloaddition of the 1-oxa-1,3-butadienes **8-14** and ethyl vinyl ether **8-12**. The difference in the activation volume of the transition structures leading to the *cis*- and *trans*-cycloadducts **8-15** and **8-16**, respectively depends on the size of the substituent R at the 2-position of the 1-oxa-1,3-butadiene. Thus, for **8-14c** with the small ester moiety a $\Delta\Delta V^{\#} = 2.4 \pm 0.2\ cm^3\ mol^{-1}$, for **8-14b** with the trifluoromethyl group a $\Delta\Delta V^{\#} = 3.8 \pm 0.1\ cm^3\ mol^{-1}$ and with the even bigger trichloromethyl moiety a $\Delta\Delta V^{\#} = 5.9 \pm 0.5\ cm^3\ mol^{-1}$ was observed.

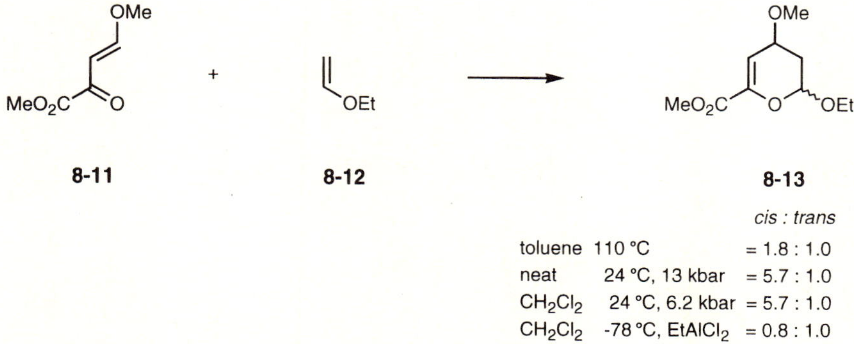

Fig. 8-4

In these systems there is a twofold advantage of applying high pressure. First, according to the remarkable large difference in activation volume the diastereoselectivity is enhanced toward high pressure at constant temperature. Secondly, also a substantial $\Delta\Delta H^{\#}$ is found causing an increase of diastereoselectivity toward lower temperature. As $\Delta V^{\#}$ is fairly large and negative, high pressure enables the cycloaddition to be run with reasonable rate even at 0 °C. Thus, the selectivity of the reaction of **8-14c** and **8-12** can be increased from 1.67 : 1.00 at 90 °C and 1 bar to 13.6 : 1.00 at 0.5 °C and 6 kbar to give **8-15a** as the major product (Fig. 8-5) [545].

The detailed analysis of the reactions has revealed that the *endo*-transition structures are influenced by high pressure to a much larger extent than the *exo*-transition structures. Thus, a pressure dependent increase in diastereoselectivity can always be expected if a high steric hindrance exists in the *endo*-transition structure [546].

In a similar way, also the cycloaddition of **8-14** to substituted enol ethers has been analysed [547]. In these investigations the correlation between steric hindrance and pressure induced diastereoselectivity is not so clear cut. However, an interesting result is the cycloaddition of **8-14b** to **8-17** to give the spiro-compound **8-18** as the major adduct, indicating that **8-17** isomerises intermediately to the *exo*-methylene enol ether which reacts faster than **8-17**. But as expected,

8-14 **8-12** **8-15** **8-16**

Pht

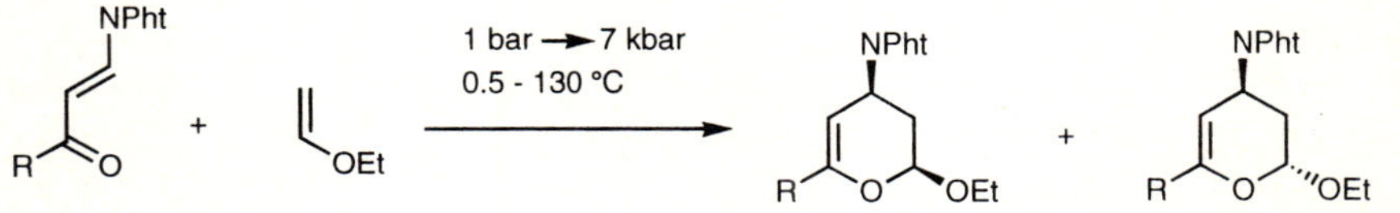

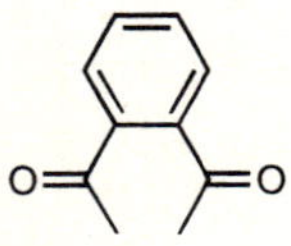

R	solvent	$\Delta\Delta V^{\#}$ [cm³/mol]	$\Delta\Delta H^{\#}$ [kJ/mol]
a: CCl_3	CH_2Cl_2	5.9 ± 0.5	8.1 ± 1.7
	n-heptane/ isodurene	5.3 ± 0.4	
b: CF_3	CH_2Cl_2	3.85 ± 0.1	8.7 ± 2.7
c: CO_2Me	CH_2Cl_2	2.4 ± 0.2	10.0 ± 0.9

Fig. 8-5

8-14b **8-17** **8-18** **8-19**

1 bar 5.2 : 1
800 bar 3.53 : 1
1600 bar 2.72 : 1

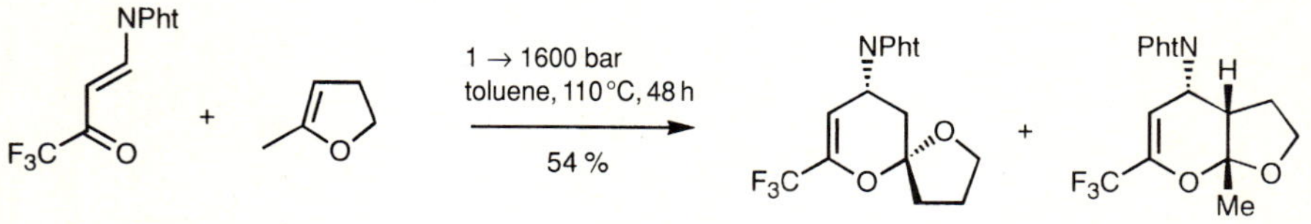

Fig. 8-6

under high pressure the formation of the annulated cycloadduct **8-19** is enhanced. Noteworthy, the isomerised enol ether could not be detected in the reaction mixture by NMR spectroscopy.

A clear correlation between the stabilisation of the *endo*-transition structure and the size of substituent at the 2-position of an 1-oxa-1,3-butadiene is again seen in the cycloaddition of the *N*-acetyl-enaminoketone **8-20** to **8-12**. As expected, the reaction of **8-20a** to give **8-21a** and **8-22a** shows only a very small $\Delta\Delta V^{\#}$, whereas with growing bulkiness of R as in **8-20b** and **8-20c** an increase of $\Delta\Delta V^{\#}$ is observed with the formation of the *trans*-cycloadduct **8-22** as the major product under high pressure. Because of the pressure effect it can clearly be deduced that **8-22** is formed via an *endo-Z-anti*-transition structure, presumably due to a strong hydrogen bond in the (*Z*)-diastereomer and a steric discrimination of the (*E*)-diastereomer of **8-20.** However, an *exo-E-anti*-transition structure would give the same product (Fig. 8-7) [548].

An effect of high pressure on the diastereoselectivity is also observed for intramolecular hetero Diels-Alder reactions as found for the cycloaddition of the benzylidene-isoxazolone **8-23** to afford the *cis*-annulated **8-24** as the major product together with the *trans*-diastereomer **8-25** (Fig. 8-8) [549]. However, the difference in activation volume with $\Delta\Delta V^{\#}=1.6\pm0.2$ cm^3 mol^{-1} is rather small. The activation volume with $\Delta V^{\#}=19.6\pm0.5$ cm^3 mol^{-1} at 343 K lies significantly below the usual values found for intermolecular cycloadditions of 1-oxa-1,3-butadienes, indicating that this reaction may be on the border line to a two-step reaction; but see also below.

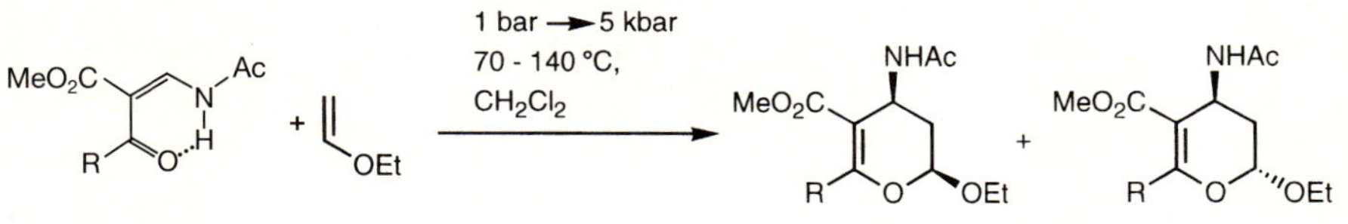

8-20 + **8-12** → (1 bar → 5 kbar, 70 - 140 °C, CH_2Cl_2) **8-21** + **8-22**

R	$\Delta\Delta V^{\#}$ [cm³/mol]	$\Delta\Delta H^{\#}$ [kJ/mol]
a: H	< 1	-
b: Et	3.8 ± 0.3	1.5 ± 0.2
c: *i*Pr	4.6 ± 0.3	2.1 ± 0.3

Fig. 8-7

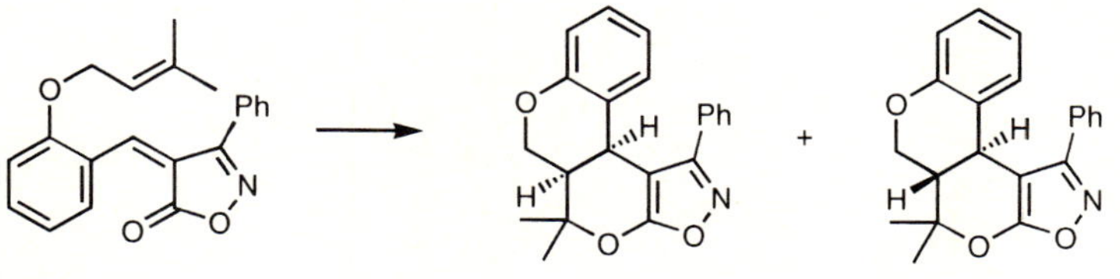

8-23 → **8-24** + **8-25**

Fig. 8-8

An influence of high pressure on the regioselectivity in hetero Diels-Alder reaction of 1-oxa-1,3-butadienes was found for the intramolecular cycloaddition of **8-26,** which led to the annulated product **8-27** and the bridged compound **8-28** (Fig. 8-9). As expected, towards higher pressure formation of the annulated product **8-27** is favoured ($\Delta\Delta V^{\#}=2.0\pm0.2$ cm^3 mol^{-1} in tetrahydrofuran). However, highly surprising is the unusual large solvent effect on the activation volume, which does not correlate with the polarity of the solvent. This must be understood as being due to a difference in solvation of the substrate in different solvents as confirmed by measurement of their molar volumes; in contrast, the molar volume of the transition structure seems to be unaffected by solvent. Thus, whenever solvent effects on organic reactions are studied by pressure-induced changes, it is recommendable to determine activation volumes and molar volumes of the substrates and products to locate the transition structure on the absolute volume scale [550].

In a similar system, namely the intermolecular cycloaddition of **8-29** to give the two enantiomers **8-30** and *ent*-**8-30** also a pressure induced increase in enantioselectivity using the Narasaka catalyst **8-31** was observed. Whereas the reaction proceeds with 4.5% *ee* at 20 °C and 1 bar, an *ee* of 20.4% was observed at 20 °C and 5 kbar (Fig. 8-10) [551].

However, the results should be taken as an exception, since in our understanding of enantioface differentiating transformations enantioselectivity is obtained by discrimination of one of the two enantiofaces of a molecule usually due to steric hindrance. Since under high pressure the sterically more crowded transition structure is preferred, a decrease in enantioselectivity should be observed under high pressure. This indeed was found for the intermolecular all-carbon Diels-Alder reaction of 2-methyl-1,3-butadiene to a crotonic acid derivative [552]. A divergent result may be obtained if the differentiation is caused by electrostatic reasons.

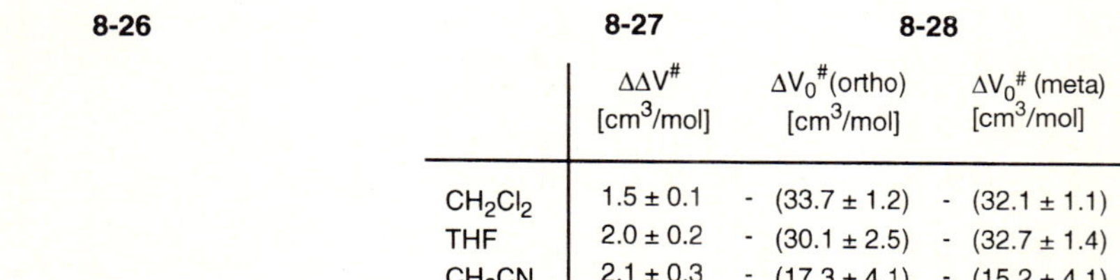

1 bar → 6 kbar
100-110 °C

8-26 **8-27** **8-28**

	$\Delta\Delta V^{\#}$ [cm^3/mol]	$\Delta V_0^{\#}$(ortho) [cm^3/mol]	$\Delta V_0^{\#}$ (meta) [cm^3/mol]
CH_2Cl_2	1.5 ± 0.1	- (33.7 ± 1.2)	- (32.1 ± 1.1)
THF	2.0 ± 0.2	- (30.1 ± 2.5)	- (32.7 ± 1.4)
CH_3CN	2.1 ± 0.3	- (17.3 ± 4.1)	- (15.2 ± 4.1)
toluene	1.3 ± 0.3	- (13.4 ± 1.5)	- (12.1 ± 1.5)

Fig. 8-9

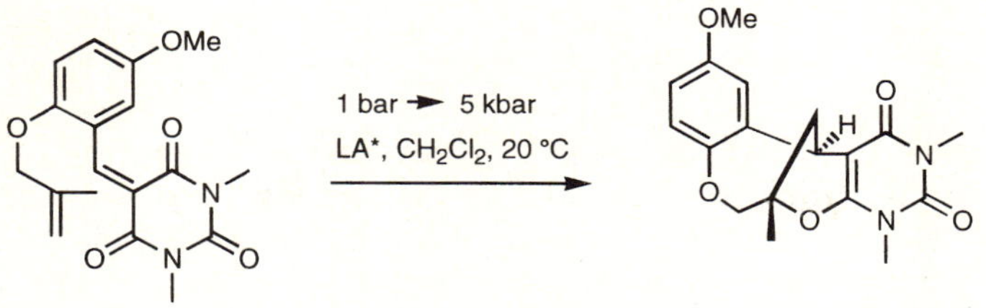

1 bar → 5 kbar
LA*, CH_2Cl_2, 20 °C

8-29 **8-30** + *ent* - **8-30**

1 bar : 4.5 % *ee*
5 kbar : 20.4 % *ee*

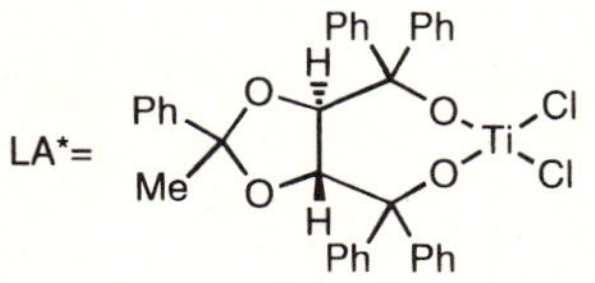

LA*= **8-31**

Fig. 8-10

Some older examples of the application of high pressure in hetero Diels-Alder reactions are found in excellent reviews and books on this topic [552–561]. In addition a few hetero Diels-Alder reaction under high pressure have already been mentioned in the foregoing chapters.

9
Novel Developments in Hetero Diels-Alder Reactions

9.1
Hetero Diels-Alder Reactions on Solid Support

A first example of a hetero Diels-Alder reaction on solid support has recently been described by us [562]. The three component domino transformation was performed by a Knoevenagel condensation of resin-linked acetoacetate **9-1** with aldehydes in the presence of piperidinium acetate at 20 °C. The obtained polymer-bound 1-oxa-1,3-butadienes were reacted with enol ethers at 60 °C in CH_2Cl_2. Final cleavage of the formed dihydropyrans from the polymer was achieved by basic transesterification with sodium methanolate. The final products **9-2** of the library were obtained in good yield and with over 90% purity without any further chromatographic purification. This demonstrates the feasibility and the great advantage of performing hetero Diels-Alder reactions on a solid support (Fig. 9-1).

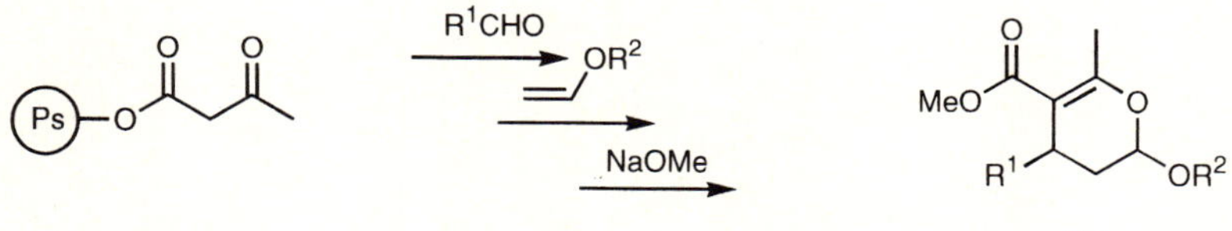

R^1CHO; OR^2 enol ether; NaOMe

9-1 **9-2**

Fig. 9-1

9.2
Use of Monoclonal Antibodies in Hetero Diels-Alder Reactions

Many examples of the use of catalytic monoclonal antibodies for a variety of organic transformations and especially for Diels-Alder reactions have been described in the last years since its discovery by Lerner and Schultz [563].

Recently, a hetero Diels-Alder reaction of an arylnitroso dienophile **9-4** and (*E*)-piperylene **9-3** to give the two regioisomeric cycloadducts **9-5** and **9-6** in the presence of a catalytic antibody has been published by Pandit and his group [564]. The most successful hapten used was the bridged compound **9-7** (Fig. 9-2).

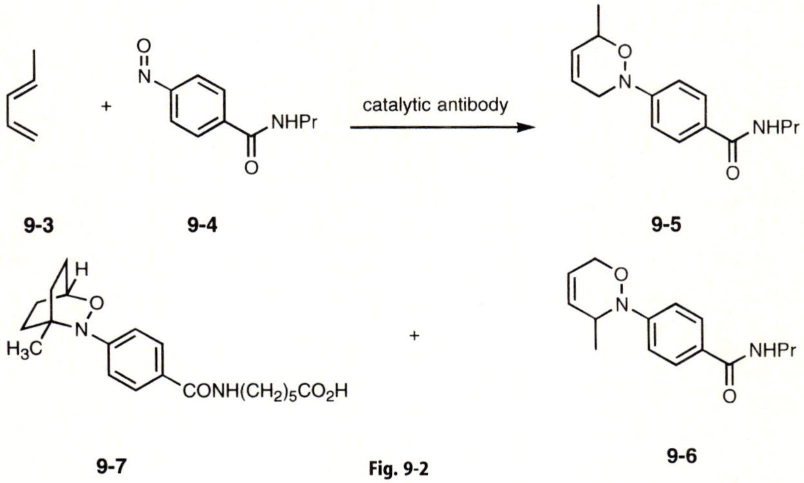

catalytic antibody

NHPr

9-3 **9-4** **9-5**

H_3C

$CONH(CH_2)_5CO_2H$

9-7 **9-6**

Fig. 9-2

There was a rate enhancement of $k_{cat}/k_{uncat}=1205$ and also a slight change in the selectivity compared to the uncatalyzed reaction (**9-5**:**9-6**:uncatalyzed =58:42; catalyzed=47:53).

A retro hetero Diels-Alder reaction to release an anthracene derivative **9-9** and nitroxyl (HNO) from the corresponding cycloadduct **9-8** by a catalytic antibody has been described by Reymond and Lerner [565]. As a haptene the acridinium salt **9-10** was used (Fig. 9-3). The antibody obtained is of great biological interest as a prodrug release system since the liberated nitroxyl is easily oxidized by the ubiquitous enzyme superoxide dismutase to give nitric oxide (NO) which acts as a chemical messenger for several important bioregulatory processes.

9.3
Hetero Diels-Alder Reactions in Aqueous Solution

Many organic reactions show a tremendous acceleration if performed in an aqueous solution [566, 567]. This has especially been shown for the Diels-Alder

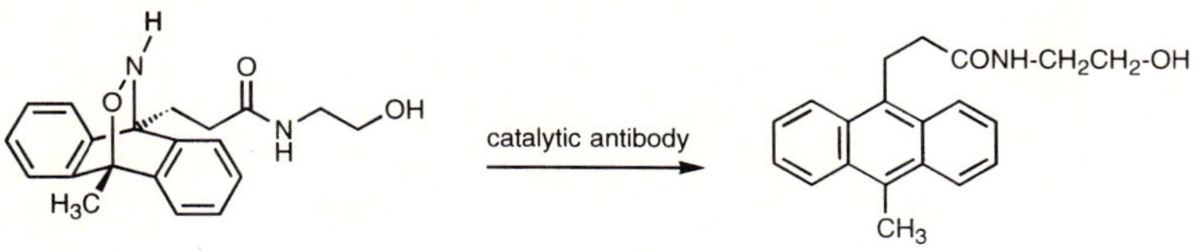

9-8 catalytic antibody 9-9 + HNO

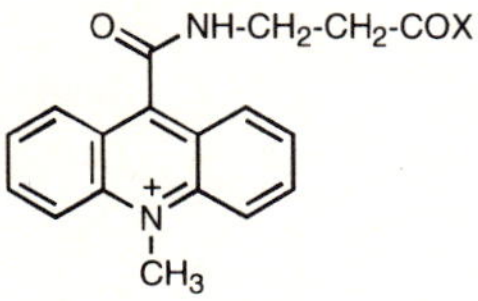

9-10

Fig. 9-3

reaction, a fact which has already been mentioned several times in this article. After the work of Grieco [568, 569] on aqueous hetero Diels-Alder reactions of iminium salts, in one of the newest examples of this type of transformation, Engberts and his group [570, 571] have compared the kinetics of the reaction of the tetrazine **9-11** and some styrols **9-12** in aqueous and organic solvents to give a dihydropyridazine **9-13** via a cycloaddition followed by a fast retro Diels-Alder reaction and a H-shift. In all cases a rate enhancement of about 100 was observed performing the reaction in water (Fig. 9-4).

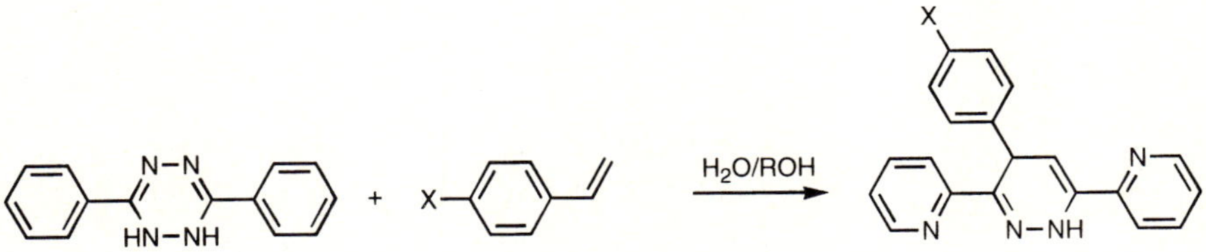

9-11 9-12 H_2O/ROH 9-13

Fig. 9-4

As another example, Lubineau [572] has shown that well available glyoxylate cycloadds to several dienes such as cyclopentadiene, cyclohexadiene and isoprene. The reaction of cyclopentadiene **9-14** and glyoxylic acid **9-15** in water at pH 0.9 is complete within 90 min at 40 °C to give the diastereomeric α-hydroxylactones **9-16** and **9-17** via a cycloadduct as the primary intermediate (Fig. 9-5).

Lately, also an enantioselective hetero Diels-Alder reaction of a butadiene and glyoxylate in water has been described. The yields and the observed selectivities were higher in water, but the effect was not very pronounced [573].

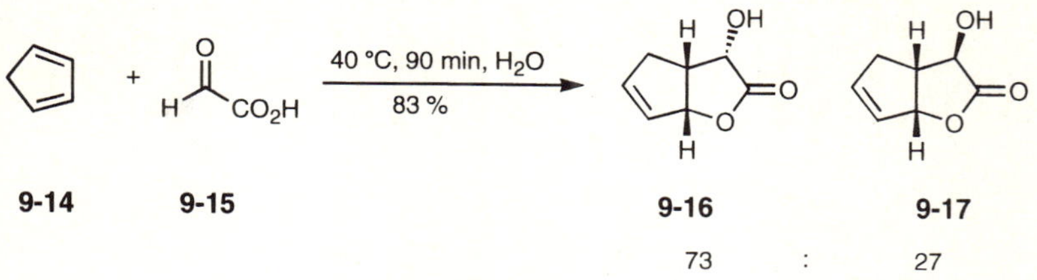

9-14 9-15 9-16 9-17

40 °C, 90 min, H_2O 83 %

73 : 27

Fig. 9-5

9.4 Microwave Activation of Hetero Diels-Alder Reactions

As already mentioned, hetero Diels-Alder reactions can be accelerated by applying Lewis acids and high pressure. However, also the application of microwaves can increase the reaction rate [574]. Thus, the usually little reactive methyl vinyl ketone **9-19** cycloadded to highly sensitive ketene acetals such as **9-18** within 10 min at 20 °C under microwave irradiation to give the dihydropyran **9-20** in 69% yield. Using other ketene acetals yields of up to 95% could be achieved (Fig. 9-6).

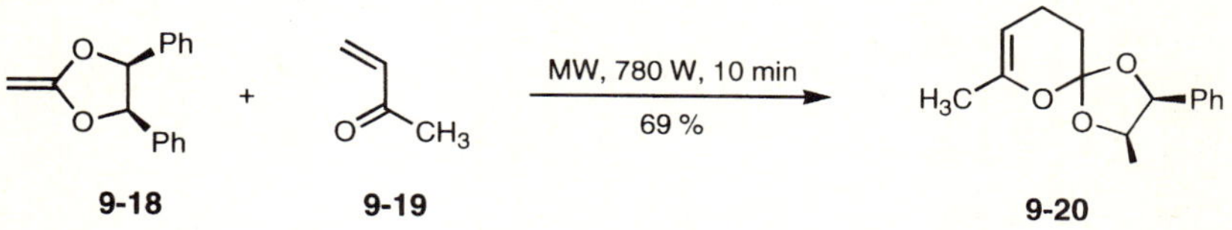

9-18 9-19 9-20

MW, 780 W, 10 min 69 %

Fig. 9-6

10 Conclusion

The multitude of hetero Diels-Alder reactions found in the literature clearly demonstrates the importance of this transformation. Thus, this type of cycloaddition is today one of the most important methods for the synthesis of heterocycles. Striking features of this method are the tremendous diversity, excellent efficiency especially in those cases where the reactive dienes and dienophiles are formed in situ, and high stereoselectivity in many cases. There is a broad scope and only little limitation. In recent years the use of Lewis acid, the development of diastereoselective and enantioselective reactions as well as the application of high pressure gave an enormous push. In addition, many of the obtained heterocycles can be transformed into acyclic compounds allowing the stereoselective preparation of e.g. amino and hydroxyl functionalized open chain compounds or even carbocycles to be of interest. Also, for the synthesis of natural products, the hetero Diels-Alder reaction is of great value. Since heterocycles, especially *N*-heterocycles, are of great pharmacological potency it can be expected that this method will have a booming importance in the field of combinatorial chemistry.

Acknowledgement. It is a particular pleasure for one of us (LFT) to warmly thank my co-workers for their commitment and their excellent performance, as reflected in the numerous scientific publications. We would like to thank Mrs. S. Williams, Miss M. Pretor and Miss A. Döring and Dipl. Chem. A. Schuffenhauer for their help in the preparation of the manuscript and the drawings. We would also like to express our gratitude to the Deutsche Forschungsgemeinschaft, the Fonds der Chemischen Industrie, the Volkswagen-Stiftung and the Ministerium für Wissenschaft und Kultur as well as to the companies BASF AG, Bayer AG, Degussa AG, Hoechst AG and Merck AG for generous support of the work described here.

References

1. Diels O, Alder L (1928) Liebigs Ann Chem 460:98
2. Hudlicky T (1996) Chem Rev 96:3
3. Tietze LF, Beifuss U (1993) Angew Chem 105:137; Angew Chem Int Ed Engl 32:131
4. Tietze LF (1996) Chem Rev 96:115
5. Denmark SE, Thorarensen A (1996) Chem Rev 96:137
6. Parsons PJ, Penkett CS, Shell AJ (1996) Chem Rev 96:195
7. Waldmann H (1994) Synthesis 535
8. Streith J, Defoin A (1994) Synthesis 1107
9. Barluenga J, Tomas M (1993) Adv. Heterocycl Chem 57:1
10. Weinreb SM (1991) In: Trost BM (ed) Comprehensive organic synthesis. Pergamon, Oxford, vol 5, p 401–449
11. Boger DL(1991) In: Trost BM (ed) Comprehensive organic synthesis. Pergamon, Oxford, vol 5, p 451–512
12. Tietze LF (1990) J Heterocyclic Chem 27:47
13. Weinreb SM, Scola PM (1989) Chem Rev 89:1525
14. Boger DL, Weinreb SM (1987) Hetero-Diels-Alder methodology in organic synthesis. Academic Press, San Diego
15. Kametani T, Hibino S, (1987) In: Advances in Heterocyclic Chemistry 42:245–333
16. Danishefsky SJ, De Ninno MP (1987) Angew Chem 99:15; Angew Chem Int Ed Engl 26:15
17. Schmidt RR (1986) Acc Chem Res 19:250
18. Boger DL (1986) Chem Rev 86:781
19. Tietze LF (1984) In: Bartmann W, Trost BM (eds) Selectivity – a goal for synthetic efficiency. Verlag Chemie, Weinheim pp 299
20. Weinreb SM, Staib RR (1982) Tetrahedron 38:3087
21. Sauer J, Sustmann R (1980) Angew Chem Int Ed Engl 19:779
22. Desimoni G, Tacconi G (1975) Chem Rev 75:651
23. Danishefsky SJ (1996) Angew Chem 108:1483; Angew Chem Int Ed Engl 35:1380
24. Laschat S (1996) Angew Chem 108:313; Angew Chem Int Ed Engl 35:289
25. Pindur U, Schneider GH (1994) Chem Soc Rev 23:409
26. Schultz PG, Lerner RA (1995) Science 269:1835
27. Oikawa H, Katayama K, Suzuki Y, Ichihara A (1995) J Chem Soc Chem Commun 1321
28. Carrol WA, Grieco PA (1993) J Am Chem Soc 115:1164
29. Bornmann WG, Kuehne ME (1992) J Org Chem 57:1752
30. Scott AI (1970) Acc Chem Res 3:151
31. Wenkert E (1962) J Am Chem Soc 84:98
32. Chapman OL, Engel MR, Springer JP, Clardy JC (1971) J Am Chem Soc 93:6696
33. Blanche F, Cameron B, Crouzet J, Debussche L, Thibaut D, Vuilhorgne M, Leeper FJ, Battersby AR (1995) Angew Chem 107:421; Angew Chem Int Ed Engl 34:383

34. Le Y, Alanine AID, Viswakarma RA, Balachandran S, Leeper FJ, Battersby AR (1994) J Chem Soc Chem Commun 2507
35. Tietze LF, Schulz G (1997) Chem J Eur 3:523
36. Boland W, Pohnert G, Maier I (1995) Angew Chem 107:1717; Angew Chem Int Ed Engl 34:1602
37. Cherkauskas JP, Klos AM, Borzilleri RM, Sisko J, Weinreb SM (1996) Tetrahedron 52:3135
38. Tietze LF, Brumby T, Pretor M, Remberg G (1988) J Org Chem 53:810
39. Review: Houk KN, Li Y, Evanseck JD (1992) Angew Chem 104: 711; Angew Chem Int Ed Engl 31:682
40. Dewar MJS, Olivella S, Stewart JJP (1986) J Am Chem Soc 108:5771
41. Tietze LF, Fennen J, Anders E (1989) Angew Chem 101:1420; Angew Chem Int Ed Engl 28:1371
42. Tietze LF, Fennen J, Schulz G, Anders E (1997) Liebigs Ann Chem: in press
43. Tietze LF, Fennen J, Geißler H, Schulz G, Anders E (1995) Liebigs Ann Chem:1681
44. Mellor JM, Merriman GD (1995) Tetrahedron 51:6115
45. Gregoire PJ, Mellor JM, Merriman GD (1995) Tetrahedron 51:6133
46. McCarrick MA, Wu YD, Houk KN (1992) J Am Chem Soc 114:1499
47. McCarrick MA, Wu YD, Houk KN (1993) J Org Chem 58:3330
48. Jursic BS, Zdravkovski Z (1994) J Phys Org Chem 7:641
49. Jursic BS (1996) J Comp Chem 17:835
50. Jursic BS, Zdravkovski Z (1995) J Org Chem 60:3163
51. Suárez D, González J, Sordo TL, Sordo JA (1994) J Org Chem 59:8058
52. Coxon JM, MCDonald DQ (1992) Tetrahedron Lett 33:3673
53. Tran Huu Dau ME, Flament JP, Lefour JM, Riche C, Grierson DS (1992) Tetrahedron Lett 33:2343
54. Tietze LF, Geissler H, Fennen J, Brumby T, Brand S, Schulz G (1994) J Org Chem 59:182
55. Tietze LF, Bratz M, Machinek R, v. Kiedrowski G (1987) J Org Chem 52:1638
56. Tietze LF, Brumby T, Brand S, Bratz M (1988) Chem Ber 121:499
57. Gresham TL, Steadman TR (1949) J Am Chem Soc 71:737
58. Dale WJ, Sisti AJ (1954) J Am Chem Soc 76:81
59. Bonjouklian R, Ruden RA (1977) J Org Chem 42:4095
60. Hosomi A, Sakata Y, Sakurai H (1985) Tetrahedron Lett 26:5175
61. Hosomi A, Otaka K, Sakurai H (1986) Tetrahedron Lett 27:2881
62. Yablonovskaya SD, Shekhtman NM, Antonova ND, Bogatkov SV, Makin SM, Zefirov NS (1970) Zh Org Khim:6871
63. Achmatowicz O, Jurczak J, Pyrek JS (1975) Rocz Chem 49:1831
64. Daniewski WM, Kubak E, Jurczak J (1985) J Org Chem 50:3963
65. Ishihara T, Shinjo H, Inoue Y, Ando T (1983) J Fluorine Chem 22:1
66. Kirby AJ, Ryder H, Matassa V (1990) J Chem Soc Perkin Trans 1:617.
67. Danishefsky SJ, Larson ER, Askin D (1982) J Am Chem Soc 104:6457
68. Bednarski M, Danishefsky SJ (1983) J Am Chem Soc 105:3716
69. Danishefsky SJ, Pearson WP, Harvey DF, Maring CJ, Springer JP (1985) J Am Chem Soc 107:1256
70. Bednarski M, Danishefsky SJ (1986) J Am Chem Soc 108:7060
71. Danishefsky SJ, Uang BJ, Quallich G (1985) J Am Chem Soc 107:1285
72. Mujica MT, Afonso MM, Galindo A, Palenzuela JA (1996) Tetrahedron 52:2167; see also Gould SJ, Eisenberg RL, Hillis LR (1991) Tetrahedron 47:7209
73. Corey EJ, Cywin CL, Roper TD (1992) Tetrahedron Lett 33:6907
74. Mikami K, Matsukawa S (1994) J Am Chem Soc 116:4077
75. Reetz MT, Raguse B, Marth CF, Hügel HM, Bach T, Fox DNA (1992) Tetrahedron 48:5731
76. Lattmann E, Coombs J, Hoffmann HMR (1996) Synthesis 171
77. Page PCB, Williams PH, Collington EW, Finch H (1987) J Chem Soc Chem Commun 756
78. Maruoka K, Saito S, Yamamoto H (1994) Synlett 439
79. Van de Weghe P, Collin J (1994) Tetrahedron Lett 35:2545
80. Forman MA, Dailey WP (1991) J Am Chem Soc 113:2761

81. Reetz MT, Gansäuer A (1993) Tetrahedron 49:6025
82. Grieco PA, Moher ED (1993) Tetrahedron Lett. 34:5567
83. Tamion R, Mineur C, Ghosez L (1995) Tetrahedron Lett 36:8977
84. Handy ST, Grieco PA, Mineur C, Ghosez L (1995) Synlett:565
85. Clive DLJ, Bergstra RJ (1990) J Org Chem 55:1786
86. Jurczak J, Tkacz M (1979) J Org Chem 44:3347
87. Schmidt RR, Abele W (1982) Angew Chem 94:298; Angew Chem Int Ed Engl 21:302
88. Bauer T, Chapius C, Kozak J, Jurczak J (1989) Helv Chim Acta 72:482
89. Bauer T, Kozak J, Chapius C, Jurczak J (1990) J Chem Soc Chem Commun 1178.
90. David S, Eustache J (1979) J Chem Soc Perkin Trans 1:2230; 2521
91. David S, Eustache J, Lubineau A (1974) J Chem Soc Perkin Trans 1:2274
92. Mulzer J, Meyer F, Buschmann J, Luger P (1995) Tetrahedron Lett 36:3503
93. Whitesell JK, Lawrence RM, Chen HH (1986) J Org Chem 51:4779
94. Lehmler HJ, Nieger M, Breitmaier E (1996) Synthesis 105
95. Arai Y, Masuda T, Masaki Y, Shiro M (1996) Tetrahedron Asymmetry 7:1199
96. Danishefsky SJ, Kobayashi S, Kerwin JF (1982) J Org Chem 47:1981
97. Garner P, Ramakanth S (1986) J Org Chem 51:2609
98. Midland MM, Afonso MM (1989) J Am Chem Soc 111:4368
99. Midland MM, Koops RW (1990) J Org Chem 55:4647; 5058
100. Savard J, Brassard P (1979) Tetrahedron Lett 51:4911
101. Reetz MT, Drewes MW, Schmitz A (1987) Angew Chem 99:1186; Angew Chem Int Ed Engl 26:1141
102. Lubineau A, Acrostanzo H, Queneau (1995) J Carb Chem 14:1307
103. Lowe RF, Stoodley RJ (1994) Tetrahedron Lett 35:6351
104. Faller JW, Smart CJ (1989) Tetrahedron Lett 30:1189
105. Maruoka K, Yamamoto H (1989) J Am Chem Soc 111:789
106. Maruoka K, Itoh T, Shirasaka T, Yamamoto H (1988) J Am Chem Soc 110:310
107. Gao Q, Maruyama T, Mouri M, Yamamoto H (1992) J Org Chem 57:1951
108. Mikami K, Motoyama Y, Terada M (1994) J Am Chem Soc 116:2812
109. Motoyama Y, Terada M, Mikami K (1995) Synlett 967
110. Gao Q, Ishihara K, Maruyama T, Mouri M, Yamamoto H (1994) Tetrahedron 50:979
111. Keck GE, Li YL, Krishnamurthy D (1995) J Org Chem 60:5998
112. Togni A (1990) Organometallics 9:3106
113. Johannsen M, Jørgensen KA (1995) J Org Chem 60:5757
114. Johannsen M, Jørgensen KA (1996) Tetrahedron 52:7321
115. Corey EJ, Cywin CL, Roper TD (1992) Tetrahedron Lett 33:6907
116. Salem L (1968) J Am Chem Soc 90:543, 553
117. Klopman G (1968) J Am Chem Soc 90:223
118. Sauer J, Sustmann R (1980) Angew Chem 92:773; Angew Chem Int Ed Engl 19:779
119. Coda AC, Desimoni G, Righetti PP, Tacconi G, Butlafava A, Martinotti-Fancitano F (1983) Tetrahedron 39:331
120. Snowden RL, Sonnay P, Ohloff G (1981) Helv Chim Acta 64:1247; Woods GF, Sanders H (1946) J Am Chem Soc 68:2483
121. Tietze LF (1974) Chem Ber 107:2491
122. Tietze LF, Glüsenkamp KH, Harms K, Remberg G, Sheldrick GM (1982) Tetrahedron Lett 23:1147
123. Tietze LF, Glüsenkamp KH, Holla W (1982) Angew Chem 94:793; Angew Chem Int Ed Engl 21:787
124. Tietze LF, Glüsenkamp KH (1983) Angew Chem 95:901; Angew Chem Int Ed Engl 22:887
125. Tietze LF, Stegelmeier H, Harms K, Brumby T (1982) Angew Chem 94:868; Angew Chem Int Ed Engl 21:863
126. Tietze LF, von Kiedrowski G, Harms K, Clegg W, Sheldrick GM (1980) Angew Chem 92:130; Angew Chem Int Ed Engl 19:134
127. Tietze LF, Brumby T, Pfeiffer T (1988) Liebigs Ann Chem 9

128. Tietze LF, Ott C, Gerke K, Buback M (1993) Angew Chem 105:1536; Angew Chem Int Ed Engl 32, 1485
129. Tietze LF, Schäfer P, unpublished results
130. a) Tietze LF, Meier H, Nutt H (1989) Chem. Ber. 122:643; b) Tietze LF, Meier H, Nutt H (1990) Liebigs Ann Chem 253
131. Mizukami S, Kihara N, Endo T (1993) Tetrahedron Lett 34:7437
132. Koser S, Hoffmann HMR (1994) Heterocycles 37:661
133. a) Weichert A, Hoffmann HMR (1991) J Org Chem 56:4098; b) Hoffmann HMR, Gassner A, Eggert U (1991) Chem Ber 124:2475
134. a) De Keyser JL, De Cock CJC, Poupaert JH, Dumont P (1988) J Org Chem 53:4859; b) Zia-Ebrahimi M, Huffmann GW (1996) Synthesis 215
135. a) Bogdanowicz-Szwed K, Palasz A (1995) Monatshefte Chem 126:1341; b) Tietze LF, Voß E, Harms K, Sheldrick GM (1985) Tetrahedron Lett 26:5273
136. a) Hojo M, Masuda R, Okada E (1990) Synthesis 347; b) Hojo M, Masuda R, Okada E (1989) Synthesis 215
137. Yamauchi M, Katayama S, Baba O, Watanabe T (1990) J Chem Soc Perkin Trans 1:3041
138. a) Bloxhan J, Dell CP (1993) J Chem Soc Perkin Trans 1:3055; b) Dell CP (1992) Tetrahedron Lett 33:699
139. Takaki K, Yamada M, Negoro K (1982) J Org Chem 47:5246
140. Apparao S, Maier ME, Schmidt RR (1987) Synthesis 896; 900
141. Tietze LF, Hartfiel U, Hübsch T, Voß E, Bogdanowicz-Szwed K, Wichmann J (1991) Liebigs Ann Chem 275
142. Tietze LF, Fennen J, Wichmann J (1992) Chem Ber 125:1507
143. Tietze LF, Hartfiel U, Hübsch T, Voß E, Wichmann J (1991) Chem. Ber 124:881
144. Boger DL, Robarge KD (1988) J Org Chem 53:3373
145. a) Sera A, Ohara M, Yamada H, Egashira E, Ueda N, Setsune J (1994) Bull Chem Soc Jpn 67:1912; b) Sera A, Ohara M, Yamada H, Egashira E, Ueda N, Setsune J (1990) Chem Lett 2043
146. a) John RA, Schmid V, Wyler H (1987) Helv Chim Acta 70:600; b) Zhuo JC, Wyler W (1993) Helv Chim Acta 76:1916; c) Zhuo JC, Wyler W, Schenk K (1995) Helv Chim Acta 78:151; d) Hashem MA, Hossain T (1995) Ind J Chem 34B:768
147. Coleman RS, Grant EB (1990) Tetrahedron Lett 31:3677
148. Pale P, Bouquant J, Chuche J, Carrupt PA, Vogel P (1994) Tetrahedron 50:8035
149. Croce PD, Ferraccioli R, La Rosa C (1993) Gass Chim Ital 123:35
150. Yamamoto Y, Suzuki H, Moro-Oka Y (1986) Chem Lett 73
151. a) Wada E, Yasuoka H, Kanemasa S (1994) Chem Lett 145; b) Wada E, Pei W, Yasuoka H, Chin U, Kanemasa S (1996) Tetrahedron 52:1205
152. Mérour J-Y, Chichereau L, Desarbre E, Gadonneix P (1996) Synthesis 519
153. Conrads M, Mattay J, Runsink J (1989) Chem Ber 122:2207
154. Desimoni G, Faita G, Righetti PP, Tacconi G (1991) Tetrahedron 47:8399
155. Desimoni G, Faita G, Righetti PP, Toma L (1990) Tetrahedron 46:7951
156. Corsico Coda A, Desimoni G, Faita G, Righetti PP, Tacconi G (1989) Tetrahedron 45:775
157. Grieco PA, Nunes JJ, Gaul MD (1990) J Am Chem Soc 112:4595
158. Desimoni G, Faita G, Righetti PP, Vietti F (1995) Heterocycles 40:817
159. Desimoni G, Faita G, Righetti PP (1995) Tetrahedron Lett 36:2855
160. Desimoni G, Faita G, Righetti PP, Jardone N (1996) personal communication
161. a) Tietze LF, Schneider C, Grote A (1996) Chem Eur J 2:139; b) Tietze LF, Schneider C, Montenbruck A (1994) Angew Chem 106:1031; Angew Chem Int Ed Engl 33: 980; c) Tietze LF, Hartfiel U (1990) unpublished results
162. Tietze LF, Schulz G (1995) Liebigs Ann Chem 1921
163. Tietze LF, Schneider C (1992) Synlett 755
164. a) Snider BB, Zhang Q (1991) J Org Chem 56:4908; b) Evans DA, Chapman KT, Bisaha J (1984) J Am Chem Soc 106:4261; (1989) 110:1238
165. Tietze LF, Brand S, Pfeiffer T, Antel J, Harms K, Sheldrick GM (1987) J Am Chem Soc 109:921

166. a) Sakaki J, Sugita Y, Sato M, Kaneko C (1991) Tetrahedron 47:6197; b) Sato M, Kano K, Kitazawa N, Hisamichi H, Kaneko C (1990) Heterocycles 31:1229
167. Sato M, Kitazawa N, Nagashima S, Kaneko C, Inoue N, Furuya (1991) Tetrahedron 47:7271
168. Sato M, Sunami S, Kaneko C, Satoh S, Furuya T (1994) Tetrahedron: Asymmetry 5:1665
169. a) Tietze LF, Beifuss U, Ruther M, Rühlmann A, Antel J, Sheldrick GM (1988) Angew Chem 100:1200; Angew Chem Int Ed Engl 27:1186; b) Tietze LF, Schulz G. (1996) Liebigs Ann 1575
170. a) Hiroi K, Umemura M, Tomikawa Y (1993) Heterocycles 35:73; b) Hiroi K, Umemura M, Fujisawa A (1992) Tetrahedron Lett 33:7161
171. a) Dondoni A, Kniezo L, Martinkova M (1994) J Chem Soc Chem Commun 963; b) Dondoni A, Kniezo L, Martinkova M, Imrich J (1997) Chem J Eur 3:424
172. Lopez JC, Lameignere E, Lukacs G (1988) J Chem Soc Chem Commun 514
173. Tietze LF, Saling P (1992) Synlett 281
174. Tietze LF, Saling P (1993) Chirality 5:329
175. Wada E, Yasuoka H, Kanemasa S (1994) Chem Lett 1637
176. Wada E, Pei W, Yasuoka H, Chin U, Kanemasa S (1996) Tetrahedron 52:1205
177. Alder K (1943) Neuere Methoden der präparativen organischen Chemie, Verlag Chemie, Weinheim/Germany
178. Padwa A, Gareau Y, Harrison B, Norman BH (1991) J Org Chem 56:2713
179. Le Coz L, Veyrat-Martin C, Wartski L, Seyden-Penne J, Bois C, Philoche-Levisalles M (1990) J Org Chem 55:4870
180. The use of lanthanide triflates as catalysts for imino-Diels-Alder reactions has been described very recently in a feature article: Kobayashi S, Ishitani H, Nagayama S (1995) Synthesis 1195
181. Waldmann H, Braun M, Dräger H (1991) Tetrahedron: Asymmetry 2:1231; for a review see: Waldmann H (1995) Synlett 133
182. Lock R, Waldmann H (1994) Liebigs Ann Chem 511
183. Waldmann H, Braun M (1991) Liebigs Ann Chem 1045
184. Waldmann H (1989) Liebigs Ann Chem 231
185. Lock R, Waldmann H (1996) Tetrahedron Lett 37:2753
186. Waldmann H, Braun M (1992) J Org Chem 57:4444
187. Waldmann H, Braun M, Weymann M, Gewehr M (1991) Synlett 881
188. Kunz H, Pfrengle W (1989) Angew. Chem 101:1041; Angew Chem Int Ed Engl 28:1067
189. Herczegh P, Kovács I, Szilágyi L, Sztaricskai F, Berecibar A, Riche C, Chiaroni A, Olesker A, Lukacs G (1995) Tetrahedron 51:2969
190. Herczegh P, Kovács I, Szilágyi L, Sztaricskai F (1994) Tetrahedron 50:13671
191. Herczegh P, Kovács I, Szilágyi L, Zsély M, Sztaricskai F (1992) Tetrahedron Lett 33:3133
192. Kündig EP, Xu LH, Romanens P, Bernardinelli G (1996) Synlett 270
193. Baldoli C, Del Buttero P, Di Ciolo M, Maiorana S, Papagni A (1996) Synlett 258
194. Bailey PD, Londesbrough DJ, Hancox TC, Heffernan JD, Holmes AB (1994) J Chem Soc Chem Comm 2543
195. Bailey PD, Brown GR, Korber F, Reed A, Wilson RD (1991) Tetrahedron Asymmetry 2:1263
196. Bailey PD, Wilson RD, Brown GR (1991) J Chem Soc Perkin Trans 1:1337
197. Ishihara K, Miyata M, Hattori K, Tada T, Yamamoto H (1994) J Am Chem Soc 116:10520
198. Hattori K, Yamamoto H (1993) Tetrahedron 49:1749
199. Hattori K, Yamamoto H (1993) Synlett 129
200. Hamley P, Helmchen G, Holmes AB, Marshall DR, MacKinnon JWM, Smith DF, Ziller JW (1992) J Chem Soc Chem Comm 786
201. McFarlane AK, Thomas G, Whiting A (1995) J Chem Soc Perkin Trans 1:2803
202. McFarlane AK, Thomas G, Whiting A (1993) Tetrahedron Lett 43:2379
203. Abraham H, Théus E, Stella L (1994) Bull Soc Chim Belg 103:361
204. Abraham H, Stella L (1992) Tetrahedron 48:9707
205. Lease TG, Shea KJ (1993) J Am Chem Soc 115:2248

206. Midland MM, Koops RW (1992) J Org Chem 57:1158
207. Barluenga J, Aznar F, Valdés C,Martín A, García-Granda S, Martín E (1993) J Am Chem Soc 115:4403; for diastereoselective aza-Diels-Alder reactions of achiral 2-amino-1,3-butadienes see: Barluenga J, Aznar F, Valdés C, Cabal M-P (1993) J Org Chem 58:3391
208. For further applications of chiral 2-amino-1,3-butadienes see: Krohn K (1993) Angew Chem 105:1651; Angew Chem Int Ed Engl 32:1582; Enders D, Meyer O (1996) Liebigs Ann 1023
209. Katagiri N, Kurimoto A, Kitano K, Nochi H, Sato H, Kaneko C (1992) Nucleic Acids Symp Ser 27:83
210. Kappe CO, Wentrup C, Kollenz G (1993) Monatsh Chem 124:1133
211. Grieco PA, Clark JD (1990) J Org Chem 55:2271
212. Grieco PA, Parker DT, Fobare WF, Ruckle R (1987) J Am Chem Soc 109:5859
213. Grieco PA, Bahsas A (1987) J Org Chem 52:5746
214. Katritzky AR, Gordeev MF (1993) J Org Chem 58:4049
215. Tietze LF, Fennen J, Geissler H, Schulz G, Anders E (1995) Liebigs Ann 1681. For further theoretical analysis of aza-Diels-Alder reactions see: Tran Huu Dau ME, Flament JP, Lefour L-M, Riche C, Grierson DS (1992) Tetrahedron Lett 33:2343; Bachrach SM, Liu M (1992) J Org Chem 57:6736
216. For an examplary transformation see: Vijn RJ, Arts HJ, Green R, Castelijns AM (1994) Synthesis 573
217. For a general discussion see ref. 14
218. Ghosez L, Serckx-Poncin B, Rivera M, Bayard P, Sainte F, Demoulin A, Hesbain-Frisque A-M, Mockel A, Munoz L, Bernard-Henriet C (1985) Lect Heterocycl Chem 8:69
219. Serckx-Poncin B, Hesbain-Frisque A-M, Ghosez L (1982) Tetrahedron Lett 23:3261
220. Del Mar Blanco M, Alonso MA, Avendaño C, Menéndez JC (1996) Tetrahedron 52:5933
221. Tapia RA, Quintanar C, Valderrama JA (1996) Heterocycles 43:447
222. Chaker L, Pautet F, Fillion H (1995) Heterocycles 41:1169
223. Brassard P, Lévesque S, (1994) Heterocycles 38:2205
224. Gómez-Bengoa E, Echavarren AM (1991) J Org Chem 56:3497
225. Kitahara Y, Kubo A (1992) Heterocycles 34:1089
226. Beaudegnies R, Ghosez L (1994) Tetrahedron Asymmetry 5:557
227. Waldner A (1989) Tetrahedron Lett 30:3061
228. Boger DL, Zhu Y (1991) Tetrahedron Lett 32:7643
229. Behforouz M, Gu Z, Cai W, Horn MA, Ahmadian M (1993) J Org Chem 58:7089
230. Pérez JM, Avendaño C, Menéndez JC (1995) Tetrahedron 51:6573
231. Boger DL, Corbett WL, Curran TT, Kasper AM (1991) J Am Chem Soc 113:1713
232. Boger DL, Nakahara S (1991) J Org Chem 56:880
233. Boger DL, Corbett WL, Wiggins JM (1990) J Org Chem 55:2999
234. Boger DL, Kasper AM (1989) J Am Chem Soc 111:1517
235. Boger DL, Corbett WL (1993) J Org Chem 58:2068
236. Uyehara T, Chiba N, Suzuki I, Yamamoto Y (1991) Tetrahedron Lett 32:4371
237. Teng M, Fowler FW (1990) J Org Chem 55:5646
238. Uyehara T, Suzuki I, Yamamoto Y (1990) Tetrahedron Lett 31:3753
239. Hwang YC, Fowler FW (1985) J Org Chem 50:2719
240. Jung ME, Choi YM (1991) J Org Chem 56:6729
241. Wojciechowski K (1993) Tetrahedron 49:7277
242. Wojciechowski (1991) Synlett 571
243. Dufour B, Motorina I, Fowler FW, Grierson DS (1994) Heterocycles 37:1455
244. Trione C, Toledo LM, Kuduk SD, Fowler FW, Grierson DS (1993) J Org Chem 58: 2075
245. Teng M, Fowler FW (1989) Tetrahedron Lett 30:2481; see also ref. 237
246. Sisti NJ, Motorina IA, Tran Huu Dau M-E, Riche C, Fowler FW, Grierson DS (1996) J Org Chem 61:3715
247. Sisti NJ, Fowler FW, Grierson DS (1991) Synlett 816
248. Sakamoto M, Nagano M, Suzuki Y, Satoh K, Tamura O (1996) Tetrahedron 52:733

249. Echavarren AM (1990) J Org Chem 55:4255; see also Nebois P, Fillion H, Benameur L (1993) Tetrahedron 49:9767
250. Gürtler CF, Steckhan E, Blechert S (1996) J Org Chem 61:4136
251. Gürtler CF, Steckhan E, Blechert S (1995) Angew Chem 107:2025; Angew Chem Int Ed Engl 34:1900
252. Bouaziz Z, Nebois P, Fillion H (1995) Tetrahedron 51:4057
253. Villacampa M, Pérez JM, Avendaño C, Menéndez JC (1994) Tetrahedron 50:10047.
254. For a recent study concerning lanthanide triflates in hetero-Diels-Alder reactions of 2-azabutadienes see: Makioka Y, Shindo T, Taniguchi Y, Takaki K, Fujiwara Y (1995) Synthesis 801
255. Mellor JM, Merriman GD (1995) Tetrahedron 51:6115
256. Mellor JM, Merriman GD, Riviere P (1991) Tetrahedron Lett 32:7103
257. Linkert F, Laschat S, Kotila S, Fox T (1996) Tetrahedron 52:955
258. Mellor JM, Merriman GD (1995) Steroids 60:693
259. Gregoire PJ, Mellor JM, Merriman GD (1991) Tetrahedron Lett 32:7099
260. Grieco PA, Bahsas A (1988) Tetrahedron Lett 29:5855
261. Barluenga J, Joglar J, González FJ, Fustero S (1990) Synlett 129, and references cited therein
262. Barluenga J, Joglar J, González FJ, Fustero S, Krüger C, Tsay Y-H (1991) Synthesis 387
263. Barluenga J, Joglar J, González FJ, Fustero S (1989) Tetrahedron Lett 30:2001
264. Barluenga J, González FJ, García-Granda S, Pérez-Carreño E (1991) J Org Chem 56:4459
265. Barluenga J, González FJ, Fustero S (1990) Tetrahedron Lett 31:397; see also ref. 261
266. Barluenga J, Pérez Carlón R, González FJ, Joglar J, López Ortiz F, Fustero S (1992) Bull Soc Chim Fr 129:566
267. Barluenga J, Pérez Carlón R, González FJ, López Ortiz F, Fustero S (1991) J Chem Soc Chem Comm 1704
268. Tietze LF, Utecht J (1992) Chem Ber 125:2259
269. For a recent study concerning intermolecular aza-Diels-Alder reactions of *N*-aryl imines see: Narasaka K, Shibata T (1993) Heterocycles 35:1039
270. Linkert F, Laschat S, Knickmeier M (1995) Liebigs Ann 985
271. Temme O, Laschat S (1995) J Chem Soc Perkin Trans 1:125
272. Linkert F, Laschat S (1994) Synlett 125
273. Laschat S, Lauterwein J (1993) J Org Chem 58:2856
274. Laschat S, Noe R, Riedel M, Krüger C (1993) Organometallics 12:3738
275. Beifuss U, Herde A, Ledderhose S (1996) J Chem Soc Chem Comm 1213. For mechanistic studies on cycloadditions of isoquinolinium salts see: Gupta RB, Franck RW (1987) J Am Chem Soc 109:5393
276. Beifuss U, Kunz O, Ledderhose S, Taraschewski M, Tonko C (1996) Synlett 34
277. Beifuss U, Ledderhose S (1995) J Chem Soc Chem Comm 2137
278. Barluenga J, Tomás M, Ballesteros A, Gotor V (1989) J Chem Soc Chem Comm 267
279. Bayard P, Ghosez L (1988) Tetrahedron Lett 29:6115
280. Bayard P, Sainte F, Beaudegnies R, Ghosez L (1988) Tetrahedron Lett 29:3799
281. Sainte F, Serckx-Poncin B, Hesbain-Frisque A-M, Ghosez L (1982) J Am Chem Soc 104:1428
282. Gouverneur V, Ghosez L (1991) Tetrahedron Lett 32:5349, for a study on analogous reactions involving achiral nitroso dienophiles see: Gouverneur V, Ghosez L (1996) Tetrahedron 52:7585
283. Pouilhès A, Langlois Y, Nshimyumukiza P, Mbiya K, Ghosez L (1993) Bull Soc Chim Fr 130:304
284. Nakahara S, Numata R, Tanaka Y, Kubo A (1995) Heterocycles 41:651
285. Bouammali B, Pautet F, Fillion H (1993) Tetrahedron 49:3125
286. Saito T, Ohkubo T, Maruyama K, Kuboki H, Motoki S (1993) Chem Lett 1127
287. Molina P, Alajarín M, Vidal A, Sánchez-Andrada P (1992) J Org Chem 57:929
288. Molina P, Arques A, Molina A (1991) Synthesis 21
289. Paquette LA, Branan BM, Rogers RD (1995) J Org Chem 60:1852

290. Barluenga J, Tomás M, Suárez-Sobrino A, López LA (1995) J Chem Soc Chem Comm 1785; for an analogous transformation with an achiral aminodiene see: Barluenga J, Aznar F, Fernández M (1995) Tetrahedron Lett 36:6551
291. Enders D, Meyer O, Raabe G, Runsink J (1994) Synthesis 66
292. Tripathy R, Franck RW, Onan KD (1988) J Am Chem Soc 110:3257
293. Thiem R, Rotscheidt K, Breitmeier E (1989) Synthesis 836
294. Kerrigan JE, McDougal PG, VanDerveer D (1993) Tetrahedron Lett 34:8055
295. Sapinall IH, Cowley PM, Stoodley RJ (1994) Tetrahedron Lett 35:3397
296. Aspinall IH, Cowley PM, Mitchell G, Stoodley RJ (1993) J Chem Soc Chem Comm 1179
297. Tsuge O, Hatta T, Yakata K, Maeda H (1994) Chem Lett 1833
298. Jenner G, Ben Salem R (1990) J Chem Soc Perkin Trans 2 1961
299. Guzmán A, Romero M, Talamás FX, Villena R, Greenhouse R, Muchowski JM (1996) J Org Chem 61:2470
300. Mazumdar SN, Mukherjee S, Sharma AK, Sengupta D, Mahajan MP (1994) Tetrahedron 50:7579
301. Mazumdar SN, Mahajan MP (1991) Tetrahedron 47:1473
302. Luthardt P, Würthwein E-U (1988) Tetrahedron Lett 29:921
303. Sain B, Singh SP, Sandhu JS (1992) Tetrahedron 48:4567
304. Sain B, Singh SP, Sandhu JS (1991) Tetrahedron Lett 32:5151
305. Barluenga J, Tomás M, Ballesteros A, López LA (1994) Heterocycles 37:1109
306. Barluenga J, Tomás M, Ballesteros A, López LA (1989) Tetrahedron Lett 30:4573
307. Tietze LF, Utecht J, unpublished results
308. South MS, Jakuboski TL (1995) Tetrahedron Lett 36:5703
309. Ferguson G, Lough AJ, Mackay D, Weeratunga G (1991) J Chem Soc Perkin Trans 1:3361
310. Ganesan A, Heathcock CH (1993) J Org Chem 58:6155
311. Stolle WAW, Frissen AE, Marcelis ATM, van der Plas HC (1992) J Org Chem 57:3000
312. Stolle WAW, Veurink JM, Marcelis ATM, van der Plas HC (1992) Tetrahedron 48:1643
313. Haider N (1992) Tetrahedron 48:7173
314. Haider N (1991) Tetrahedron 47:3959
315. Nesi R, Giomi D, Turchi S, Falai A (1995) J Chem Soc Chem Comm 2201
316. Nesi R, Giomi D, Turchi S, Paoli P (1994) Tetrahedron 50:9189
317. Boger DL, Kochanny MJ (1994) J Org Chem 59:4950
318. Boger DL, Honda T, Menezes RF, Colletti SL, Dang Q, Yang W (1994) J Am Chem Soc 116:82
319. Boger DL, Honda T (1994) J Am Chem Soc 116:5647
320. Boger DL, Honda T, Menezes RF, Colletti SL (1994) J Am Chem Soc 116:5631
321. Boger DL, Honda T, Dang Q (1994) J Am Chem Soc 116:5619
322. Boger DL, Menezes RF, Dang Q (1992) J Org Chem 57:4333
323. Boger DL, Menezes RF, Honda T (1993) Angew Chem 105:310; Angew Chem Int Ed Engl 32:273
324. Boger DL, Dang Q (1992) J Org Chem 57:1630
325. Li J-H, Snyder JK (1994) Tetrahedron Lett 35:1485
326. Benson SC, Li J-H, Snyder JK (1992) J Org Chem 57:5285
327. Taylor EC, Macor JE, French LG (1991) J Org Chem 56:1807
328. Barlow MG, Sibous L, Suliman NNE, Tipping AE (1996) J Chem Soc Perkin Trans 1:519
329. Taylor EC, French LG (1989) J Org Chem 54:1245
330. Boger DL, Zhang M (1991) J Am Chem Soc 113:4230
331. Boger DL, Baldino CM (1993) J Am Chem Soc 115:11418
332. Streith J, Defoin A (1996) Synlett 189
333. Gouverneur V, Ghosez L (1990) Tetrahedron Asymmetry 1:363, see also ref. 282
334. Defoin A, Brouillard-Poichet A, Streith J (1992) Helv Chim Acta 75:109
335. Defoin A, Brouillard-Poichet A, Streith J (1991) Helv Chim Acta 74:103
336. Braun H, Felber H, Kresze G, Schmidtchen FP, Prewo R, Vasella A (1993) Liebigs Ann Chem 261
337. Felber H, Kresze G, Prewo R, Vasella A (1986) Helv Chim Acta 69:1137

338. Werbitzky O, Klier K, Felber H (1990) Liebigs Ann Chem 267
339. Braun H, Burger W, Kresze G, Schmidtchen FP (1990) Tetrahedron Asymmetry 1:403
340. Schürrle K, Beier B, Piepersberg W (1991) J Chem Soc Perkin Trans 1:2407
341. Defoin A, Sarazin H, Strehler C, Streith J (1994) Tetrahedron Lett 35:5653
342. Kirby GW, Nazeer M (1993) J Chem Soc Perkin Trans 1:1397
343. Miller A, Procter G (1990) Tetrahedron Lett 31:1043; Miller A, Procter G (1990) Tetrahedron Lett 31:1041
344. Shustov GV, Rauk A (1995) Tetrahedron Lett 36:5449
345. Defoin A, Pires J, Tissot I, Tschamber T, Bur D, Zehnder M, Streith J (1991) Tetrahedron Asymmetry 2:1209
346. Behr J-B, Defoin A, Pires J, Streith J, Macko L, Zehder M (1996) Tetrahedron 52:3283
347. Defoin A, Pires J, Streith J (1991) Synlett 417
348. Hussain A, Wyatt PB (1993) Tetrahedron 49:2123
349. Fritz H, Henlin JM, Riesen A, Tschamber T, Zehnder M, Streith J (1988) Helv Chim Acta 71:822
350. Tschamber T, Craig CJ, Muller M, Streith J (1996) Tetrahedron 52:6201
351. Behr J-B, Defoin A, Mahmood N, Streith J (1995) Helv Chim Acta 78:1166
352. Behr J-B, Defoin A, Streith J (1994) Heterocycles 37:747
353. Ghosh A, Miller MJ (1995) Tetrahedron Lett 36:6399
354. Ghosh A, Ritter AR, Miller MJ (1995) J Org Chem 60:5808
355. Ritter AR, Miller MJ (1994) J Org Chem 59:4602
356. Naruse M, Aoyagi S, Kibayashi C (1996) J Chem Soc Perkin Trans 1:1113
357. Naruse M, Aoyagi S, Kibayashi C (1994) Tetrahedron Lett 35:9213
358. Quadrelli P, Gamba Invernizzi A, Caramella P (1996) Tetrahedron Lett 37:1909
359. Soulié J, Betzer J-F, Muller B, Lallemand J-Y (1995) Tetrahedron Lett 36:9485
360. Naylor (née Bathgate) A, Howarth N, Malpass JR (1993) Tetrahedron 49:451
361. Kefalas P, Grierson DS (1993) Tetrahedron Lett 34:3555
362. Pindur U, Kim M-H (1989) Tetrahedron 45:6427
363. Gilchrist TL (1983) Chem Soc Rev 12:53
364. Reissig H-U, Hippeli C, Arnold T (1990) Chem Ber 123:2403
365. Jursic BS, Zdravkovski Z (1995) J Org Chem 60:3163
366. Hippeli C, Reissig H-U (1987) Synthesis 77
367. Hippeli C, Basso N, Dammast F, Reissig H-U (1990) Synthesis 26
368. Zimmer R, Reissig H-U (1992) J Org Chem 57:339
369. Henning R, Lerch U, Urbach H (1989) Synthesis 265
370. Nakanishi S, Higuchi M, Flood T (1986) J Chem Soc Chem Comm 30
371. Hippeli C, Reissig H-U (1990) Liebigs Ann Chem 217
372. Zimmer R, Reissig H-U (1991) Liebigs Ann Chem 553
373. Zimmer R, Reissig H-U (1988) Angew Chem 100:1576; Angew Chem Int Ed Engl 27:1518
374. Zimmer R, Reissig H-U (1989) Synthesis 908
375. Zimmer R, Reissig H-U, Lindner HJ (1992) Liebigs Ann Chem 621
376. Li JP, Newlander KA, Yellin TO (1988) Synthesis 73
377. Plate R, Nivard RJF, Ottenhijm HCJ (1987) J Chem Soc Perkin Trans 1:2473
378. Arnold T, Reissig H-U (1990) Synlett 514
379. Arnold T, Orschel B, Reissig H-U (1992) Angew Chem 104:1084; Angew. Chem Int Ed Engl 31:1033
380. Paulini K, Reissig H-U, Rademacher P (1995) J Prakt Chem 337:209
381. Paulini K, Gerold A, Reissig H-U (1995) Liebigs Ann 667
382. Hofmann B, Reissig H-U (1994) Chem Ber 127:2337
383. Paulini K, Reissig H-U (1994) Chem Ber 127:685
384. Angermann J, Homann K, Reissig H-U, Zimmer R (1995) Synlett 1014
385. Hippeli C, Reissig H-U (1990) Liebigs Ann Chem 475
386. Hippeli C, Zimmer R, Reissig H-U (1990) Liebigs Ann Chem 469
387. Chinchilla R, Bäckvall J-E (1992) Tetrahedron Lett 33:5641
388. Denmark SE, Kesler BS, Moon Y-C (1992) J Org Chem 57:4912

389. Denmark SE, Schnute ME (1994) J Org Chem 59:4576
390. Denmark SE, Marcin LR (1993) J Org Chem 58:3857
391. Denmark SE, Marcin LR (1995) J Org Chem 60:3221
392. Denmark SE, Schnute ME, Senanayake CBW (1993) J Org Chem 58:1859
393. Denmark SE, Stolle A, Dixon JA, Guagnano V (1995) J Am Chem Soc 117:2100
394. Denmark SE, Schnute ME, Marcin LR, Thorarensen A (1995) J Org Chem 60:3205
395. Papchikhin A, Agback P, Plavec J, Chattopadhyaya J (1993) J Org Chem 58:2874
396. Avalos M, Babiano R, Cintas P, Higes FJ, Jiménez JL, Palacios JC, Silva MA (1996) J Org Chem 61:1880
397. Kirby GW, Lochead AW, Sheldrake GN (1984) J Chem Soc Chem Comm 922
398. Baldein JE, Lopez RCG (1983) Tetrahedron 39:1487
399. Blandon CM, Ferguson IEG, Kirby GW, Lochead AW (1985) J Chem Soc Perkin Trans 1:1541
400. Hasserodt J, Pritzkow H, Sundermeyer W (1995) Liebigs Ann 95
401. Choi SS-M, Kirby GW (1991) J Chem Soc Perkin Trans 1:3225
402. Vedejs E, Eberlein TH, Wilde RG (1988) J Org Chem 53:2220
403. Segi M, Takahashi M, Nakajima T, Suga S (1989) Synth Comm 19:2431
404. Vedejs E, Stults JS, Wilde RG (1988) J Am Chem Soc 110:5452
405. Takahashi T, Kurose N, Koizumi T (1993) Heterocycles 36:1601
406. Bonini BF, Mazzanti G, Zani P, Maccagnani G (1988) J Chem Soc Chem Comm 365
407. Vedejs E, Stults JS (1988) J Org Chem 53:2226
408. Revesz L, Siegel RA, Buescher H-H, Marko M, Maurer R, Meigel H (1990) Helv Chim Acta 73:326
409. Pinto IL, Buckle DR, Rami HK, Smith DG (1992) Tetrahedron Lett 33:7597
410. Adam D, Freer AA, Isaacs NW, Kirby GW, Littlejohn A, Rahman MS (1992) J Chem Soc Perkin Trans 1:1261
411. Tromm P, Heimgartner H (1988) Helv Chim Acta 71:2071
412. Karakasa T, Motoki S (1978) J Org Chem 43:4147, Karakasa T, Motoki S (1979) J Org Chem 44:4151
413. Karasaka T, Yamaguchi H, Motoki S (1980) J Org Chem 45:927
414. Barnish IT, Fishwick CWG, Hill DR (1991) Tetrahedron Lett 32:405
415. Gabbutt CD, Hepworth JD, Heron BM (1992) J Chem Soc Perkin Trans 1:2603
416. Baruah PD, Mukherjee S, Mahajan MP (1990) Tetrahedron 46:1951
417. Greif D, Mulst M, Weissenfels M (1987) Synthesis 456
418. Blitzke T, Greif D, Kempe R, Pink M, Pulst M, Sicker D, Wilde H (1994) J Prakt Chem 336:163
419. Baruah PD, Nongkynrih I, Mahajan MP (1989) Synthesis 631
420. Murase M, Nishino N, Nara N, Nakanishi Y, Tobinaga S (1994) Heterocycles 37:725
421. Barnish IT, Fishwick CWG, Hill DR, Szantay C Jr (1989) Tetrahedron 45:6771
422. Barnish IT, Fishwick CWG, Hill DR, Szantay C Jr (1989) Tetrahedron 45:7879
423. Barnish IT, Fishwick CWG, Hill DR, Szantay C Jr (1989) Tetrahedron Lett 30:4449
424. Saito T, Nagashima M, Karakasa T, Motoki S (1992) J Chem Soc Chem Comm 411
425. Saito T, Nagashima M, Karasaka T, Motoki S (1990) J Chem Soc Chem Comm 1665
426. Moriyama S, Karakasa T, Inoue T, Kurashima K, Satsumabayashi S, Saito T (1996) Synlett 72
427. Saito T, Kimura H, Sakamaki K, Karakasa T, Moriyama S (1996) J Chem Soc Chem Comm 811
428. Motoki S, Saito T, Karakasa T, Matsushita T, Furono E (1992) J Chem Soc Perkin Trans 1:2943
429. Motoki S, Saito T, Karakasa T, Kato H, Matsushita T, Hayashibe S (1991) J Chem Soc Perkin Trans 1:2281
430. Saito T, Karakasa T, Fujii H, Furono E, Suda H, Kobayashi K (1994) J Chem Soc Perkin Trans 1:1359
431. Bell AS, Fishwick CWG, Reed JE (1996) Tetrahedron Lett 37:123
432. Ohno M, Kojima S, Eguchi S (1995) J Chem Soc Chem Comm 565

433. Ohno M, Kojima S, Shirakawa Y, Eguchi S (1995) Tetrahedron Lett 36:6899; for similar oxa-Diels-Alder reaction leading to pyran-fused fullerenes see: Ohno M, Azuma T, Eguchi S (1993) Chem Lett 1833
434. Meier H, Eckes H-L, Niedermann H-P, Kolshorn H (1987) Angew Chem 99:1040; Angew Chem Int Ed Engl 26:1046
435. Jacob D, Niedermann H-P, Meier H (1986) Tetrahedron Lett 27:5703
436. Saito T, Shizuta T, Kikuchi H, Nakawaga J, Hirotsu K, Ohmura H, Motoki S (1994) Synthesis 727
437. Moriyama S, Mochizuki T, Ohshima Y, Saito T (1994) Bull Chem Soc Jpn 67:2876
438. Beifuss U, Gehm H, Taraschewski M (1996) Synlett 396
439. Barluenga J, Tomás M, Ballesteros A, López LA (1989) Tetrahedron Lett 30:6923
440. Barluenga J, Tomás M, Ballesteros A, López LA (1995) Synthesis 985
441. Barluenga J, Tomás M, Ballesteros A, López LA (1991) Synlett 93
442. Barluenga J, Tomás M, Ballesteros A, López LA (1991) J Org Chem 56:5680
443. Marchand A, Mauger D, Guingant A, Pradère J-P (1995) Tetrahedron Asymmetry 6:853
444. Chehna M, Pradère J-P, Quiniou H, Le Botlan D, Toupet L (1989) Phosphorus, Sulfur, and Silicon 42:15
445. Swindell CS, Tao M (1993) J Org Chem 58:5889
446. Nair V, Kumar S (1996) J Chem Soc Perkin Trans 1:443
447. Nair V, Kumar S (1994) J Chem Soc Chem Comm 1341
448. Dondoni A, Fogagnolo M, Mastellari A, Pedrini P (1986) Tetrahedron Lett 27:3915
449. Hartke K, Lindenblatt T (1990) Synthesis 281
450. Capozzi G, Franck RW, Mattioli M, Menichetti S, Nativi C, Valle G (1995) J Org Chem 60:6416
451. Capozzi G, Dios A, Franck RW, Geer A, Marzabadi C, Menichetti S, Nativi C, Tamarez M (1996) Angew Chem 108:805; Angew Chem Int Ed Engl 35:777
452. Leblanc Y, Fitzsimmons BJ, Springer JP, Rokach J (1989) J Am Chem Soc 111:2995
453. Toepfer A, Schmidt RR (1993) Carbohydr Res 247:159
454. Weinreb SM (1988) Acc Chem Res 21:313
455. Garigipati RS, Cordoba R, Parvez M, Weinreb SM (1986) Tetrahedron 42:2979
456. Bell SI, Parvez M, Weinreb SM (1991) J Org Chem 56:373, and literature cited therein
457. Turos E, Parvez M, Garigipati RS, Weinreb SM (1988) J Org Chem 53:1116
458. Bell SI, Weinreb SM (1988) Tetrahedron Lett 29:4233
459. Hamada T, Sato H, Hikota M, Yonemitsu O (1989) Tetrahedron Lett 30:6405
460. Tornus I, Schaumann E (1996) Tetrahedron 52:725
461. Deguin B, Vogel P (1993) Tetrahedron Lett 34:6269
462. Deguin B, Vogel P (1993) Helv Chim Acta 76:2250
463. Deguin B, Vogel P (1992) J Am Chem Soc 114:9210
464. Suárez D, González J, Sordo TL, Sordo JA (1994) J Org Chem 59:8058
465. Suárez D, Sordo TL, Sordo JA (1994) J Am Chem Soc 116:763
466. Suárez D, Assfeld X, González J, Ruiz-López MF, Sordo TL, Sordo JA (1994) J Chem Soc Chem Comm 1683
467. Gilchrist TL, Wood JE (1992) J Chem Soc Perkin Trans 1:9
468. Steliou K, Gareau Y, Milot G, Salama P (1990) J Am Chem Soc 112:7819
469. Regitz M, Binger P (1988) Angew Chem 100:1541; Angew Chem Int Ed Engl 27:1484
470. Märkl G, Dorsch S (1995) Tetrahedron Lett 36:3839; b) Märkl G, Dorfmeister G (1987) Tetrahedron Lett 28:1093
471. Driess M, Pritzkow H, Rell S, Winkler U (1996) Organometallics 15:1845
472. Sekiguchi A, Maruki I, Ebata K, Kabuto C, Sakurai H (1991) J Chem Soc Chem Comm 341
473. Paetzold P, Kiesgen J, Krahé K, Meier H-U, Boese R (1991) Z Naturforschung (B) 46:853
474. Segi M, Kato M, Nakajima T (1991) Tetrahedron Lett 7427
475. Bauer T, Jezewski A, Jurczak J (1996) Tetrahedron Asymmetry 7:1405
476. Golebiowski A, Jacobsson U, Chmielewski M, Jurczak J (1987) Tetrahedron 43:599

477. Lubineau A, Queneau Y (1995) J Carbohydr Chem 14:1295
478. Tietze LF, Montenbruck A, Schneider C (1994) Synlett 509; see also ref. 161 and 163
479. De Gaudenzi L, Apparao S, Schmidt RR (1990) Tetrahedron 46:277
480. Haag-Zeino B, Schmidt RR (1990) Liebigs Ann Chem 1197
481. Dujardin G, Rossignol S, Brown E (1996) Tetrahedron Lett 37:4007
482. Tietze LF, Bachmann J, Wichmann J, Burkhardt O (1994) Synthesis 1185.
483. Martin SF, Clark CW, Corbett JW (1995) J Org Chem 60:3236
484. Martin SF, Benage B, Geraci LS, Hunter JE, Mortimore M (1991) J Am Chem Soc 113:6161
485. Martin SF, Benage B, Unter JE (1988) J Am Chem Soc 110:5925
486. Takano S, Ohkawa S, Tamori S, Satoh S, Ogasawara K (1988) J Chem Soc Chem Comm:189
487. Takano S, Satoh S, Ogasawara K (1988) J Chem Soc Chem Comm 59
488. Takano S (1987) Pure Appl Chem 59:353
489. Tietze LF, Meier H, Nutt H (1990) Liebigs Ann Chem 253
490. Tietze LF, Bärtels C (1991) Liebigs Ann Chem 155
491. Tietze LF, Denzer H, Holdgrün X, Neumann M (1987) Angew. Chem 99:1309; Angew Chem Int Ed Engl 26:1295
492. Tietze LF, Wölfling J, Schneider G (1991) Chem Ber 124:591
493. Tietze LF, Beifuss U, Lökös M, Rischer M, Göhrt A, Sheldrick GM (1990) Angew Chem 102:545; Angew. Chem Int Ed Engl. 29:527
494. Tietze LF, Schneider C (1991) J Org Chem 56:2476
495. Ireland RE, Armstrong JD, III, Lebreton J, Meissner RS, Rizzacasa MA (1993) J Am Chem Soc 115:7152
496. Snider BB, Lu Q (1996) J Org Chem 61:2839
497. Snider BB, Lu Q (1994) J Org Chem 59:8065
498. Burke SD, Piscopio AD, Buchanan JL (1988) Tetrahedron Lett 29:2757
499. Burke SD, Piscopio AD, Kort ME, Matulenko MA, Parker MH, Armistead DM, Shankaran K (1994) J Org Chem 59:332
500. Burke SD, Armistead DM, Shankaran K (1986) Tetrahedron Lett 27:6295
501. Inoue T, Inoue S, Sato K (1990) Bull Chem Soc Jpn 63:1647
502. Tietze LF, von Kiedrowski G, Berger B (1982) Angew Chem 94:222; Angew Chem Int Ed Engl 21:221
503. Bayles R, Flynn AP, Galt RHB, Kirby S, Turner RW (1988) Tetrahedron Lett 29:6341
504. Vogt K, Schmidt RR (1988) Tetrahedron 44:3271
505. Pfrengle W, Kunz H (1989) J Org Chem 54:4261
506. Kaufman MD, Grieco PA (1994) J Org Chem 59:7197
507. Carroll WA, Grieco PA (1993) J Am Chem Soc 115:1164
508. Hamada T, Zenkoh T, Sato H, Yonemitsu O (1991) Tetrahedron Lett 32:1649
509. Trova MP, Mc Gee, KF Jr (1995) Tetrahedron 51:5951
510. Boger DL, Cassidy KC, Nakahara S (1993) J Am Chem Soc. 115:10733
511. Boger DL, Hüter O, Mbiya K, Zhang M (1995) J Am Chem Soc 117:11839
512. Echavarren AM, Stille JK (1988) J Am Chem Soc 110:4051
513. Heathcock CH, Hansen MM, Ruggeri RB, Kath JC (1992) J Org Chem 57:2544; see also: Heathcock CH (1992) Angew Chem 104:675; Angew Chem Int Ed Engl 31:665
514. Heathcock CH, Kath JC, Ruggeri RB (1995) J Org Chem 60:1120
515. Heathcock CH, Ruggeri RB, McClure KF (1992) J Org Chem 57:2585
516. Heathcock CH, Stafford JA, Clark DL (1992) J Org Chem 57:2575
517. Heathcock CH, Stafford JA (1992) J Org Chem 57:2566
518. Gupta RB, Franck RW (1989) J Am Chem Soc 111:7668
519. King SB, Ganem B (1994) J Am Chem Soc 116:562
520. Defoin A, Sarazin H, Streith J (1996) Helv Chim Acta 79:560
521. Defoin A, Sarazin H, Streith J (1995) Synlett 1187
522. Hudlicky T, Olivo HF (1991) Tetrahedron Lett 32:6077
523. Hudlicky T, Olivo HF (1992) J Am Chem Soc 114:9694

524. Martin SF, Tso H-H (1993) Heterocycles 35:85
525. Martin SF, Hartmann M, Josey JA (1992) Tetrahedron Lett 33:3583
526. Benbow JW, McClure KF, Danishefsky SJ (1993) J Am Chem Soc 115:12305
527. McClure KF, Danishefsky SJ (1993) J Am Chem Soc 115:6094
528. Benbow JW, Schulte GK, Danishefsky SJ (1992) Angew Chem 104:934; Angew Chem Int Ed Engl 31:915
529. McClure KF, Danishefski SJ (1991) J Org Chem 56:850
530. Burkholder TP, Fuchs PL (1990) J Am Chem Soc 112:9601
531. Keck GE, Romer DR (1993) J Org Chem 58:6083
532. Naruse M, Aoyagi S, Kibayashi C (1994) J Org Chem 59:1358
533. Shishido Y, Kibayashi C (1992) J Org Chem 57:2876
534. Aoyagi S, Shishido Y, Kibayashi C (1991) Tetrahedron Lett 32:4325
535. Denmark SE, Thorarensen A, Middleton DS (1995) J Org Chem 60:3574. For a related approach to (–)-hastanecine see: Denmark SE, Thorarensen A (1994) J Org Chem 59:5672
536. Vedejs E, Wittenberger SJ (1990) J Am Chem Soc 112:4357
537. Vedejs E, Reid JG, Rodgers JD, Wittenberger SJ (1990) J Am Chem Soc 112:4351
538. Vedejs E, Rodgers JD, Wittenberger SJ (1988) J Am Chem Soc 110:4822
539. Dujardin G, Maudet M, Brown E (1994) Tetrahedron Lett 35:8619
540. Jurczak J, Golebiowski A, Raczko J (1988) Tetrahedron Lett 29:5975
541. Bauer T, Jurczak J (1992) Polish J Chem 66:1999
542. Achmatowicz O, Bialecka-Florianczyk E (1990) Tetrahedron 46:5317
543. Vandenput DAL, Scheeren HW (1995) Tetrahedron 51:8383
544. Boger DL, Robarge KD (1988) J Org Chem 53:3373
545. a) Tietze LF, Hübsch T, Voss E, Buback M, Tost W (1988) J Am Chem Soc 110:4065; b) Buback M, Tost W, Hübsch T, Voss E, Tietze LF (1989) Chem Ber 122:1179
546. Tietze LF, Hübsch T, Oelze J, Ott C, Tost W, Wörner G, Buback M (1992) Chem Ber 125:2249
547. Buback M, Kuchta G, Niklaus A, Henrich M, Rothert I, Tietze LF (1996) Liebigs Ann 1151
548. Tietze LF, Hübsch T, Ott C, Kuchta G, Buback M (1995) Liebigs Ann 1
549. Buback M, Abeln J, Hübsch T, Ott C, Tietze LF (1995) Liebigs Ann 9
550. Buback M, Gerke K, Ott C, Tietze LF (1994) Chem Ber 127:2241
551. Tietze LF, Ott C, Gerke K, Buback M (1993) Angew Chem 105:1536; Angew Chem Int Ed Engl 32, 1485
552. Tietze LF, Ott C, Frey U (1996) Liebigs Ann 63
553. Winter R, Jonas J (1993) High pressure chemistry, biochemistry and materials science. Kluwer, Dordrecht
554. Matsumoto K, Acheson RM (1991) Organic synthesis at high pressure. Wiley-Interscience, New York
555. Jurczak J, Baranowski B (1989) High pressure chemical synthesis. Elsevier, Amsterdam
556. Le Noble WJ (1988) Organic high pressure chemistry. Elsevier, Amsterdam
557. Isaacs NS (1981) Liquid phase high pressure chemistry. Wiley, Chichester
558. Isaacs NS (1991) Tetrahedron 47:8463
559. Klärner FG (1989) Chem Unserer Zeit 23:53
560. Van Eldik R, Asano T, le Noble WJ (1989) Chem Rev 89:549
561. a) Matsumoto K, Sera A, Uchida T (1985) Synthesis 1; b) Matsumoto K, Sera A (1985) Synthesis 999
562. Tietze LF, Hippe T, Steinmetz A (1996) Synlett 1043
563. Schultz PG, Lerner RA (1995) Science 269:1835.
564. Meekel AP, Resmini M, Pandit UK (1995) J Chem Soc Chem Commun:571
565. Bahr N, Güller R, Reymond JL, Lerner RA (1996) J Am Chem Soc 118:3550
566. Review: Lubineau A, Augé J, Queneau Y (1994) Synthesis 741
567. Review: Li CJ (1993) Chem Rev 93:2023
568. Grieco PA, Larsen SD (1985) J Am Chem Soc 107:1768

569. Grieco PA, Larsen SD, Fobare WF (1986) Tetrahedron Lett 27:1975.
570. Wijnen JW, Zavarise S, Engberts JBFN (1996) J Org Chem 61:2001
571. Engberts JBFN (1993) Angew Chem 105:1610; Angew Chem Int Ed Engl 32:1545
572. Lubineau A, Augé J, Grand E, Lubin N (1994) Tetrahedron 50:10265
573. Mikami K, Kotera O, Motoyama Y, Sakaguchi H (1995) Synlett 975
574. Diaz-Ortiz A, Diez-Barra E, de la Hoz A, Prieto P, Moreno A (1994) J Chem Soc Perkin Trans 1:3595

Tandem Processes of Metallo Carbenoids for the Synthesis of Azapolycycles

Albert Padwa

Department of Chemistry, Emory University Atlanta, Georgia 30322 USA
E-mail: chemap@emory.edu

Mesoionic compounds have proven to be valuable intermediates in organic chemistry from both physical and synthetic perspectives. These substances contain a masked 1,3-dipole within their framework and are therefore willing participants in 1,3-dipolar cycloadditions. Our interest in the chemistry of mesoionic dipoles stems from studies in our laboratory dealing with the rhodium(II)-catalyzed reactions of α-diazo carbonyl compounds in the presence of various heteroatoms. The isomünchnone class of mesoionics is easily generated from the Rh(II)-catalyzed reaction of α-diazo imides and readily undergoes cycloaddition with both electron-rich and electron-deficient dipolarophiles. This article compiles our findings in the general area of cascade reactions of isomünchnones together with relevant work from other laboratories.

Keywords

Diazo, carbenoid, cyclization, dipolar-cycloaddition, carbonyl ylide, rhodium, tandem

Contents

Topics in Current Chemistry, Vol. 189

1 Introduction

Tandem reactions are among the most powerful strategic tools available to the synthetic organic chemist because they rapidly increase the complexity of a substrate while at the same time making economical use of available functional groups [1–12]. Over the past several years, transition-metal based cascade reactions have gained particular importance in organic synthesis since they allow simultaneous formation of more than one bond in a single synthetic operation with high stereoselectivity [13–20]. The Rh(II)-catalyzed tandem cyclization-cycloaddition reaction of α-diazo ketones [21] has been developed in these laboratories as a general approach to nitrogen-containing polycyclic compounds [22]. Our primary purpose in writing this review is to highlight the importance of tandem processes of metallo carbenoids for azapolycyclic synthesis. It is the intent of this article to broadly define the boundaries of our present knowledge in this field. Such an overview will put into perspective what has been accomplished and hopefully provide impetus for further investigation of this general approach.

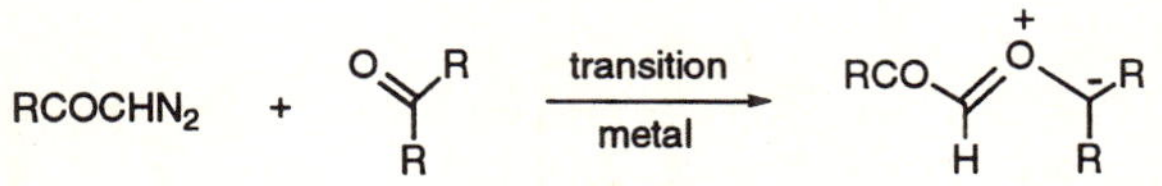

Scheme 1

2 Isomünchnone Cycloadditions

The 1,3-oxazolium-4-oxides (isomünchnones) are readily obtained through the transition metal-catalyzed cyclization of a suitable α-diazoimide [23]. This type of mesoionic oxazolium ylide corresponds to the cyclic equivalent of a carbonyl ylide and readily undergoes 1,3-dipolar cycloaddition. The first successful preparation and isolation of an isomünchnone induced by a transition metal process was described by Ibata and Hamaguchi in 1974 [24]. They observed that when diazoimide **1** was heated in the presence of a catalytic amount of $Cu(acac)_2$, a red crystalline material precipitated from the reaction mixture. The red solid was assigned as isomünchnone **4** on the basis of its spectral data and elemental analysis. Mesoionic ylide **4** was found to be air-stable for several weeks and its overall stability was attributed to its dipolar aromatic resonance structure. Formation of the isomünchnone ring can be rationalized by initial generation of a metallo-carbenoid species which is then followed by cyclization onto the neighboring carbonyl oxygen to form the dipole [25].

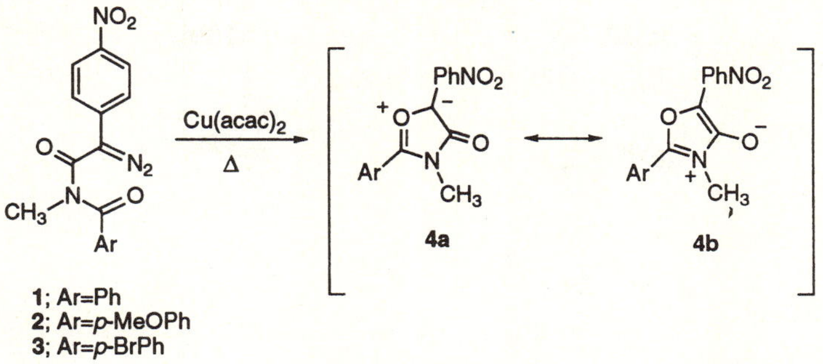

1; Ar=Ph
2; Ar=p-MeOPh
3; Ar=p-BrPh

Scheme 2

Our research group [26–28] as well as Maier's [29–30] has independently utilized the Rh(II)-catalyzed reaction of diazoimides as a method for generating isomünchnones. The starting diazoimides are readily constructed by acetoacylation [31] or malonyl-acylation [32] of the corresponding amides followed by standard diazo transfer techniques [33]. Intramolecular trapping of the rhodium carbenoid by the lone pair of electrons of the neighboring carbonyl group leads to the desired mesoionic system **6**. Both groups have shown that these reactive species can be trapped with dipolarophiles to give cycloadducts in high yield.

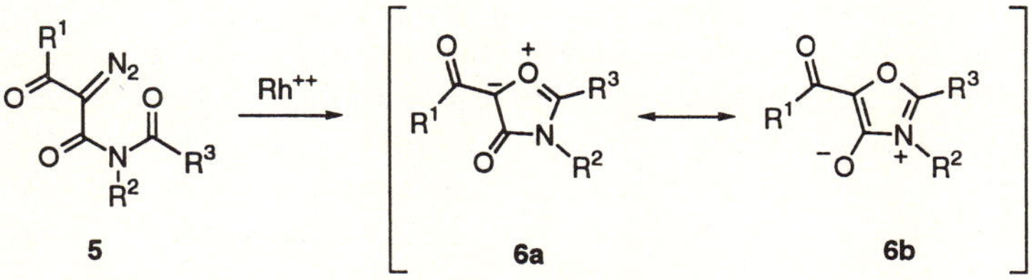

Scheme 3

Ibata was the first to show that the "masked" carbonyl ylide embedded within the isomünchnone framework would readily undergo 1,3-dipolar cycloaddition with various dipolarophiles [34]. The isolable isomünchnone **4** was observed to react with dimethyl fumarate to produce cycloadduct **7** which possesses the 7-oxa-2-azabicyclo[2.2.1]heptane skeleton. When the reaction of **1** was carried out using catalytic amounts of Cu(acac)$_2$ in the presence of various dipolarophiles, smooth dipolar cycloaddition was observed to occur giving mixtures of *endo* and *exo* isomers. In most cases, the *exo* isomers were favored. All of Ibata's isomünchnone cycloadditions contain aromatic substituent groups, presumably selected to facilitate dipole formation. The synthetic utility of the cycloaddition reaction is diminished, however, because of the low reactivity of the aromatic substituents toward further manipulation.

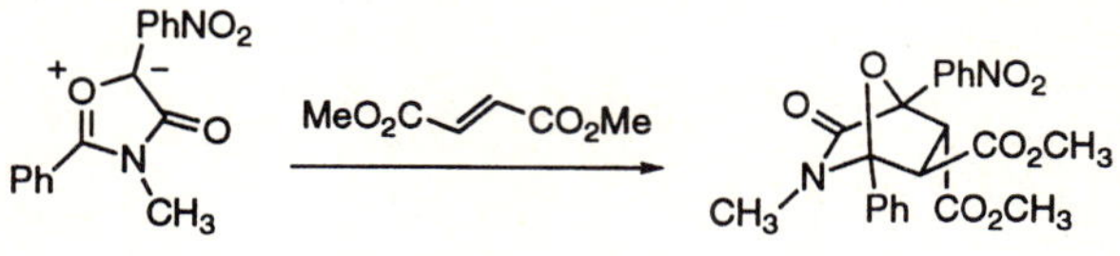

Scheme 4

Several years ago our research group became interested in using the dipolar cycloaddition of isomünchnones for the construction of a variety of alkaloid systems [26–28].Since little was known about the interaction of rhodium carbenoids with amido carbonyl groups, we sought to answer several questions: (1) would a nucleophilic amide or imide functionality cyclize more or less efficiently than a keto group to form a carbonyl ylide? (2) would the reactive diazo ketone in the presence of an activated π-bond be subject to cycloaddition across the diazo group producing a pyrazoline cycloadduct? and (3) would the given propensity for metal carbenoids to undergo addition and C-H insertion reactions be competitive with isomünchnone formation? [35]. To help answer these questions, the Rh(II)-catalyzed reactions of cyclic diazoimides **8-11** were investigated. When diazoimide **9** (n = 1) was treated with $Rh_2(OAc)_4$ in benzene (80 °C), the initially formed rhodium carbenoid cyclized onto the adjacent imide carbonyl group to generate isomünchnone **12**. This reactive species readily underwent 1,3-dipolar cycloaddition with *N*-phenyl-maleimide to give cycloadduct **14** (n = 1) as a 1.2:1 mixture of *exo*/*endo* isomers in 78% yield. No evidence of β-lactam formation, derived from competitive C-H insertion, was observed in the crude reaction mixture [35]. The ring size was reduced to a four-membered ring (**8**; n=0) and enlarged to a six- (**10**; n = 2) and seven- membered ring (**11**; n = 3). In all cases, high yields (i.e. 70–90%) of the expected cycloadducts (**13**, **15**, and **16**) were obtained. Interestingly, the cyclic cases where n = 1 and n = 3 (i.e. **9** and **11**) showed little *exo*/*endo* selectivity, but the cases of n = 0 and n = 2 (**8** and **10**) resulted in a single diastereomer.

The results obtained clearly demonstrated that the initially formed rhodium carbenoid prefers to cyclize onto the adjacent imide carbonyl group to form an isomünchnone rather than undergo C–H insertion. The explanation proposed to rationalize this result is that the preferred rhodium carbenoid conformer **17** is

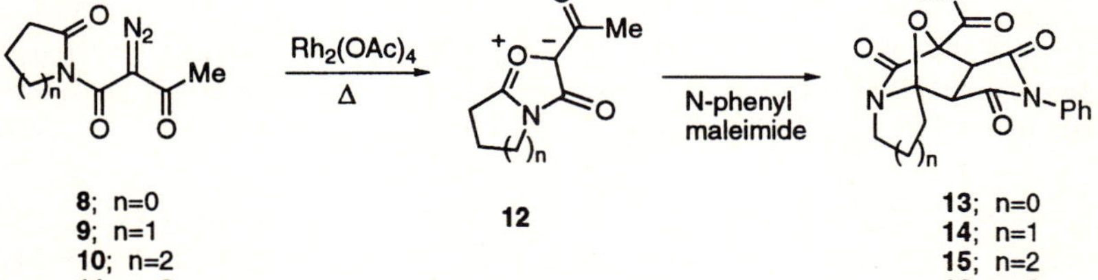

Scheme 5

the one which avoids unfavorable dipole repulsion between the two carbonyl groups of the imide (i.e. **18**). The conformational rigidity imposed by the cyclic imide ring was demonstrated to be inconsequential for carbonyl ylide formation. This was shown by carrying out the tandem cyclization-cycloaddition sequence with acyclic imides **19** and **20**. Both substrates readily reacted with *N*-phenylmaleimide to give diastereomeric mixtures of cycloadducts **21** and **22** in good yield. Again, no products derived from C–H insertion into the *N*-substituents were observed.

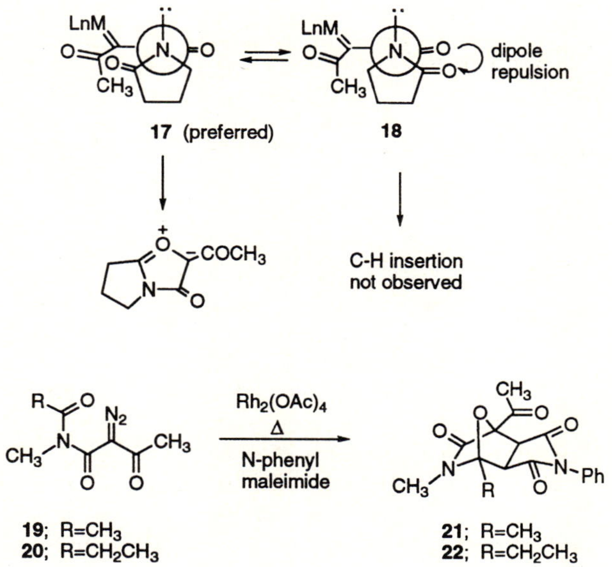

17 (preferred)

18

C-H insertion not observed

$Rh_2(OAc)_4$, Δ, N-phenyl maleimide

19; R=CH_3
20; R=CH_2CH_3

21; R=CH_3
22; R=CH_2CH_3

Scheme 6

When diazoimide **19** (or **20**) was deacetylated [36] and the resulting diazoamide **23** (or **24**) was subjected to rhodium(II) acetate, the yield of the corresponding cycloadduct (i.e. **25** or **26**) was significantly diminished. One explanation for this different reactivity is the inherent decrease in electrophilic character conferred upon the intermediate rhodium carbenoid when the diazo carbon bears a hydrogen atom rather than an acetyl group. This decrease in electrophilicity may alter the rate of carbenoid attack on the remote carbonyl group to the point where alternative reactions can occur. Another possible explanation to account for the diminished reactivity is that the preferred conformation of the intermediate rhodium carbenoid may not be the one that results in carbonyl ylide formation [35].

Unsymmetrical dipolarophiles were found to undergo intermolecular cycloaddition with isomünchnones with high regioselectivity [27]. For example, the decomposition of diazoimide **9** with $Rh_2(OAc)_4$ in the presence of methyl vinyl ketone resulted in the formation of two products identified as **27** and **28** in 27%

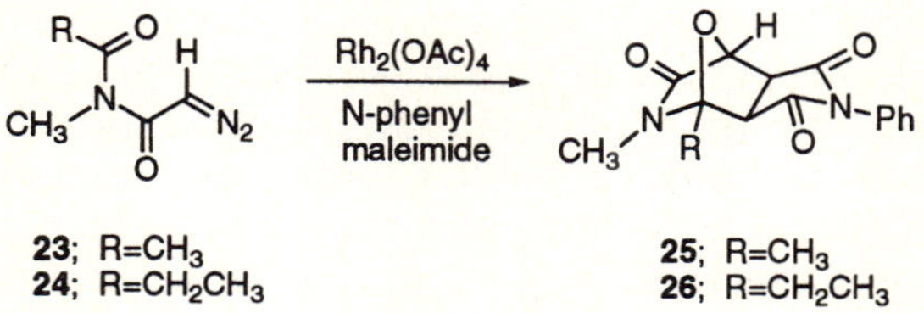

23; R=CH_3
24; R=CH_2CH_3

$Rh_2(OAc)_4$ / N-phenyl maleimide

25; R=CH_3
26; R=CH_2CH_3

Scheme 7

and 44% yield, respectively. The regiochemical outcome is consistent with FMO considerations [27]. Prolonged heating of cycloadduct **27** afforded the bicyclic lactam **28**. This rearrangement presumably occurs through nitrogen lone pair-assisted opening of the oxygen bridge of **27** to give an acyl iminium ion which then undergoes proton loss.

The first example of a bimolecular 1,3-dipolar cycloaddition between an isomünchnone and an electron-rich dipolarophile was reported by our group several years ago [27]. The reaction of diethyl ketene acetal and isomünchnone **9** gave cycloadduct **29** in high yield. Again, only one regioisomer was obtained and the regiochemistry encountered is consistent with cycloaddition involving the HOMO of diethyl ketene acetal and the LUMO of isomünchnone **12** (n = 1).

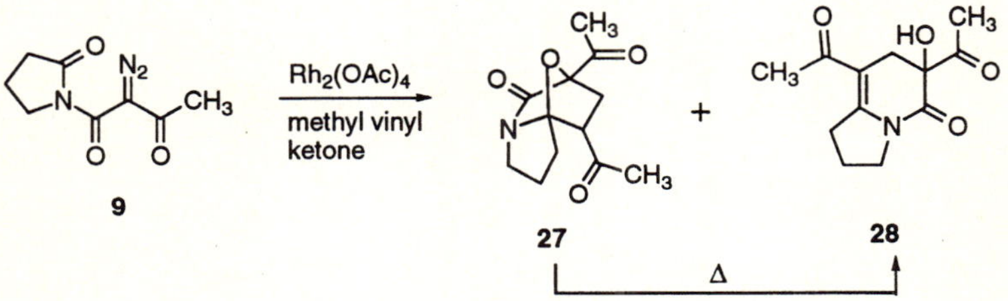

9 → $Rh_2(OAc)_4$ / methyl vinyl ketone → 27 + 28; 27 → Δ → 28

Scheme 8

9 → $Rh_2(OAc)_4$ / $CH_2=C(OEt)_2$ → 29

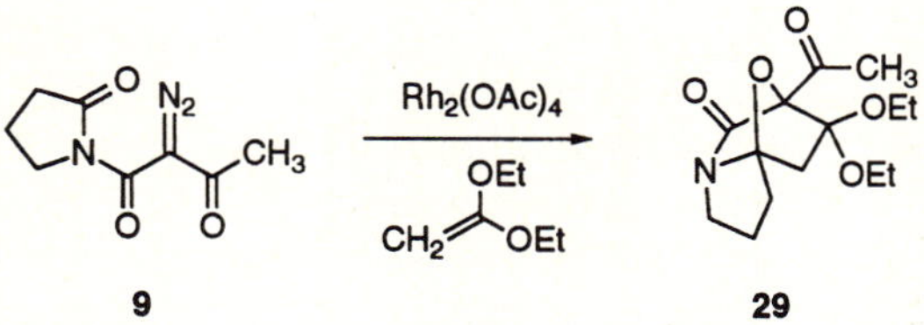

Scheme 9

3
Intramolecular Isomünchnone Cycloadditions

An interesting example of an intramolecular 1,3-dipolar cycloaddition of an isomünchnone with an unactivated alkene to produce a complex polycyclic compound in one step has been reported [26–30]. The isomünchnones derived from the $Rh_2(OAc)_4$-catalyzed reaction of acyclic diazoimides **30**–**34** were found to undergo facile cycloaddition onto the tethered π-bond to provide polycyclic adducts **35**–**39**. A notable feature of this cycloaddition is that only one diastereomer is formed. The relative stereochemistry of cycloadduct **39** was determined by X-ray crystallography [29]. This confirmed the fact that addition of the olefin took place *endo* with regard to the isomünchnone dipole. Only low yields of cycloadducts were observed when the deacylated diazoimides were subjected to the cyclization-cycloaddition reaction [29]. This result indicates that the reactivity of the 1,3-dipole is significantly diminished in the absence of the electron-withdrawing acyl group and that alternative pathways then become competitive.

This methodology was further extended, leading to an increase in complexity of the resulting polyheterocyclic systems, by employing a series of cyclic diazoimides [28]. Treatment of cyclic diazoimides **40**–**42** with $Rh_2(OAc)_4$ led to good yields of cycloadducts **43**–**45**. Only one diastereomer was produced in each cycloaddition. Once again, the stereochemical outcome is the result of an *endo* cyclization of the π-bond onto the isomünchnone dipole and this was confirmed by an X-ray crystallographic analysis of cycloadduct **43** [28].

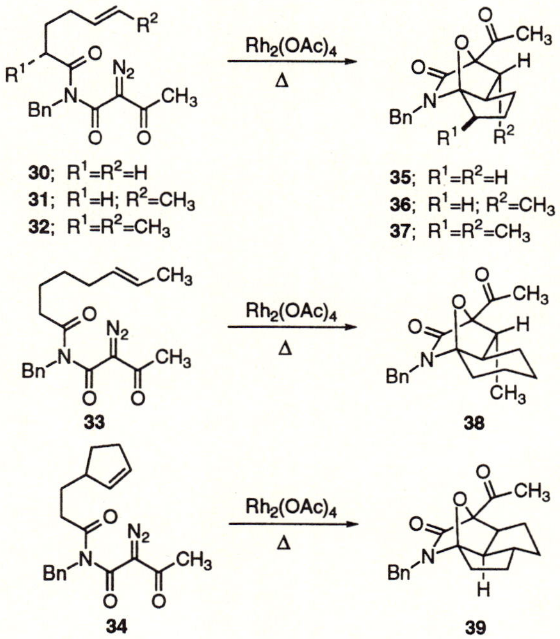

$Rh_2(OAc)_4$, Δ

30; $R^1=R^2=H$
31; $R^1=H$; $R^2=CH_3$
32; $R^1=R^2=CH_3$

35; $R^1=R^2=H$
36; $R^1=H$; $R^2=CH_3$
37; $R^1=R^2=CH_3$

33 → **38**

34 → **39**

Scheme 10

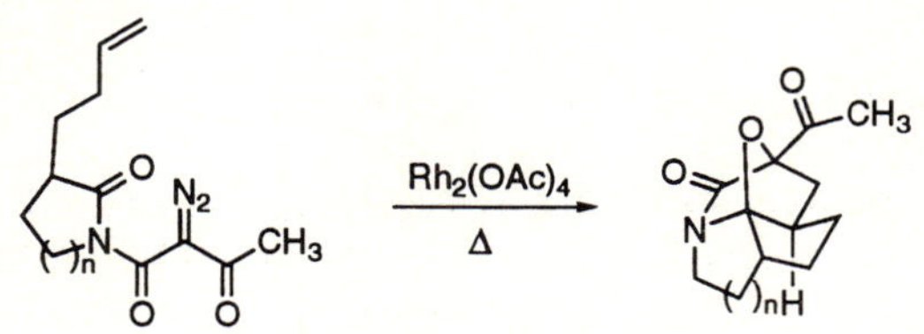

40; n=1
41; n=2
42; n=3

43; n=1 (88%)
44; n=2 (86%)
45; n=3 (83%)

Scheme 11

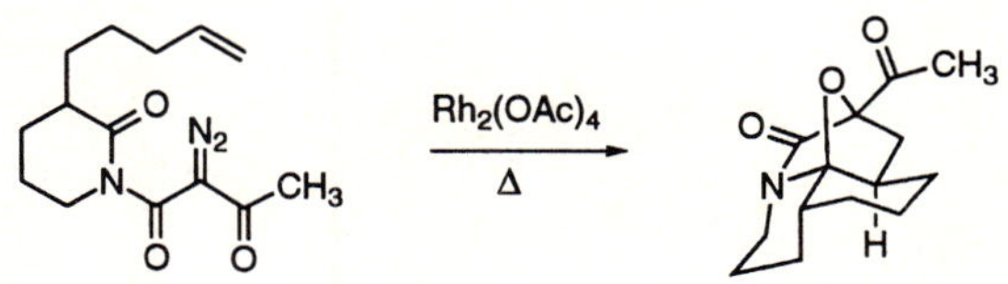

46

47

Scheme 12

Lengthening the alkenyl tether by one carbon atom was observed to have no effect on the ability of the isomünchnone to cycloadd across the olefinic 1-bond. This was shown in a study of the cycloaddition behavior of diazoimide **46** which afforded cycloadduct **47** in 86% yield as a single diastereomer [28].

The generality of the method was further demonstrated by synthesizing cyclic diazoimides **48** and **49** in which the alkenyl tether was placed alpha to the nitrogen atom [28]. When these diazoimides were treated with a catalytic amount of $Rh_2(OAc)_4$, the tandem cyclization-cycloaddition process gave polycycles **50** and **51** in 69% and 76% yield, respectively. With both of these systems, the length of the alkenyl tether proved to be crucial for the intramolecular cycloaddition reaction across the isomünchnone dipole. Only when the tether was a butenyl group was cycloaddition observed. If the length of the tether was increased or decreased by one methylene unit, no products derived from intramolecular cycloaddition were encountered [28].

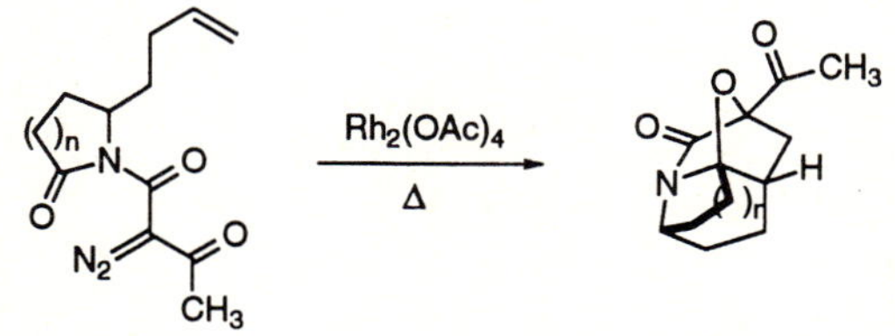

48; n=1
49; n=2

50; n=1
51; n=2

Scheme 13

4 Cyclization-Cycloaddition-Cationic π-Cyclization Reactions

The 1,3-dipolar cycloaddition of isomünchnones derived from α-diazoimides of type **52** provides a uniquely functionalized cycloadduct (i.e. **53**) containing a "masked" *N*-acyliminium ion. By incorporating an internal nucleophile on the tether, annulation of the original dipolar cycloadduct **53** would allow the construction of a more complex nitrogen heterocyclic system, particularly B-ring homologues of the erythrinane family of alkaloids [37]. By starting from simple acyclic diazoimides **52**, our research group has established a tandem cyclization-cycloaddition-cationic 1-cyclization protocol as a method for the construction of complex nitrogen poly-heterocycles of type **54**.

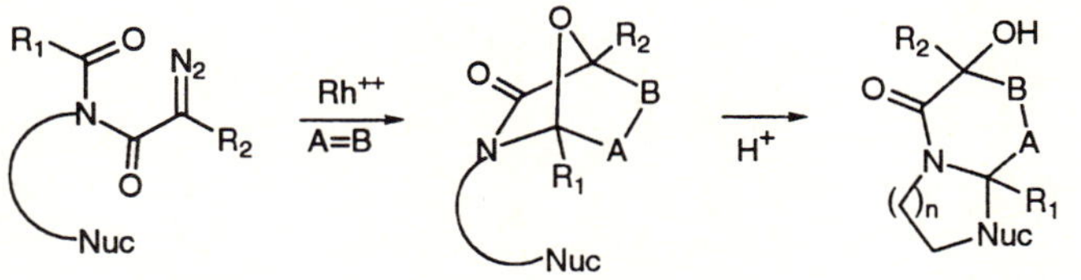

Scheme 14

The first example of such a process involved the treatment of diazoimides **55**, **56** and **57** with a catalytic quantity of rhodium(II) perfluorobutyrate in CH_2Cl_2 at 25 °C. The cycloadducts **58** (98%), **59** (95%), and **60** (90%) were produced. Formation of the *endo*-cycloadduct with respect to the carbonyl ylide dipole in these cycloadditions is in full accord with molecular mechanics calculations which show a large energy difference between the two diastereomers. When the individual cycloadducts were exposed to $BF_3 \cdot OEt_2$ (2 equiv) in CH_2Cl_2 at 0 °C, the cyclized products **61** (97%), **62** (95%), and **63** (85%) were isolated as single diastereomers. The *cis* stereochemistry of the A/B ring junction for **61-63** was assigned by analogy to similar erythrinane products obtained via a Mondon-enamide type cyclization [38–40] and was unequivocally verified by an X-ray crystal analysis of all three cycloadducts. In all three cases the *anti* stereochemical relationship is still maintained between the hydroxyl stereocenter (from the oxygen bridge) and the bridgehead proton (R_2=H) or methyl (R_2=CH_3) group.

When the dipolar cycloadduct **65**, derived from the unsubstituted alkenyl diazoimide **64**, was exposed to $BF_3 \cdot OEt_2$, the resulting cyclized product **66** (90%) was identified as the all *syn* tetracyclic lactam **66** by X-ray crystal analysis. Thus, in contrast to the other three systems, the bridgehead proton of **66** lies *syn* to the hydroxyl stereocenter of the original cycloadduct.

It is assumed that the intermediate *N*-acyliminium ions formed from the Lewis acid-assisted ring opening of the isomünchnone cycloadducts undergo rapid proton loss to produce tetra-substituted enamides. In the case of **65**, this process is clearly evident as witnessed by the stereochemical outcome observed

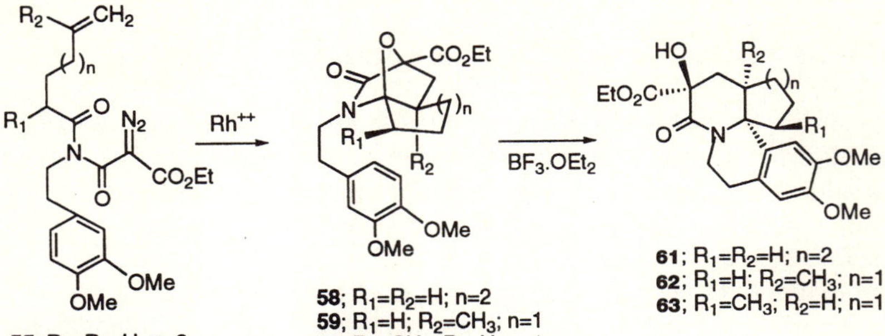

55; R_1=R_2=H; n=2
56; R_1=H; R_2=CH_3; n=1
57; R_1=CH_3; R_2=H; n=1

58; R_1=R_2=H; n=2
59; R_1=H; R_2=CH_3; n=1
60; R_1=CH_3; R_2=H; n=1

61; R_1=R_2=H; n=2
62; R_1=H; R_2=CH_3; n=1
63; R_1=CH_3; R_2=H; n=1

Scheme 15

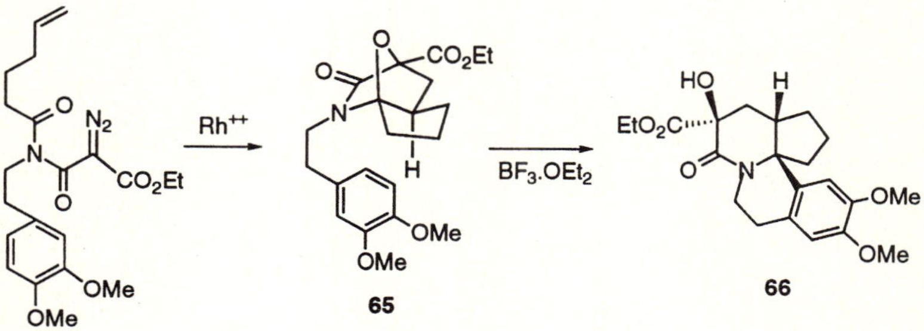

Scheme 16

in product **66**. Loss of the bridgehead proton H_A in **67** (dihedral angle 90° with respect to the *N*-acyliminium ion π-bond) is fast relative to π-cyclization. Intramolecular axial reprotonation of enamide **69** from the β-face generates the diastereomeric iminium ion **70** which then undergoes intramolecular cationic 1-cyclization from the least sterically congested face to give the observed all *syn* isomer **66**. Molecular mechanics calculations show that the *cis* A/B ring fusion in **66** is 4.6 kcal favored over the *trans* diastereomer and presumably some of this thermodynamic energy difference is reflected in the transition state for cyclization. The additional methyl group present in the related 6/5 cycloadduct (i.e. **68**) promotes loss of the proton adjacent to it and this results in the formation of enamide **71**. Stereoselective reprotonation from the least congested α-face regenerates **68** which is trapped intramolecularly by the aromatic nucleus. Cyclization always occurs from the least hindered side as has already been established by Mondon and coworkers [38]. Cationic cyclizations of this type are known to be governed by steric control [41]. In the case of cycloadduct **59**, the bridgehead proton does not exist and thus deprotonation can occur in only one direction. Apparently the initially formed iminium ion, derived from **58** (i.e. **67b**; n=2),

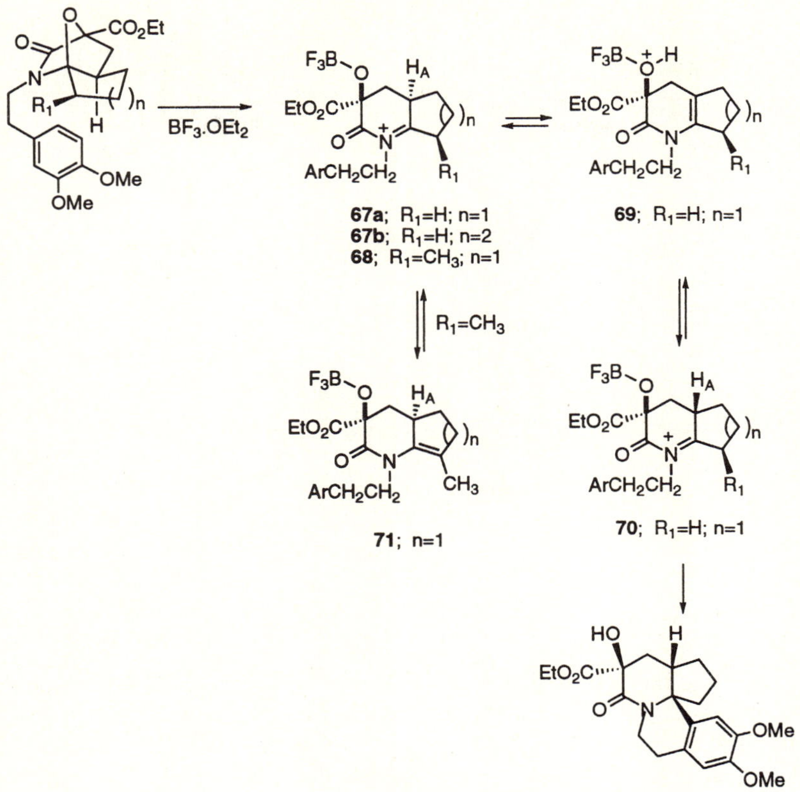

$BF_3.OEt_2$

67a; R_1=H; n=1
67b; R_1=H; n=2
68; R_1=CH_3; n=1

69; R_1=H; n=1

R_1=CH_3

71; n=1

70; R_1=H; n=1

66

Scheme 17

undergoes fast π-cyclization prior to proton loss. In this case, the deprotonation step is significantly slower than in the 6/5 system due to the larger dihedral angle (113°) proton H_a and the π-system of the *N*-acyliminium ion. The stereochemical outcome in **61** is the result of a stereoelectronic preference of the aromatic ring of the *N*-acyliminium ion for axial attack from the least hindered side.

Two additional systems which illustrate the scope and variety of π-systems which can be employed in this tandem process are outlined below. The Rh(II)-catalyzed reaction of diazoimide **72** gave rise to a transient bicyclic adduct that was not isolable, as it underwent rapid ring opening to give the conjugated indenyl enamide **73** (85%). Exposure of **73** to $BF_3 \cdot OEt_2$ in CH_2Cl_2 at 40 °C resulted in a 3:1-mixture of diastereomeric tetracyclic lactams **74** in 88% yield thereby demonstrating that tethered alkenes can also be utilized in the third step of these cascade reactions. Another substitution variation that was also investigated corresponded to the placement of an indolyl tether on the amide nitrogen. Thus, treatment of diazoimide **75** with $Rh_2(pfb)_4$ gave cycloadduct **76** (98%) which was readily converted to **77** as a single diastereomer (in 60% isolated yield). The stereochemical assignment is based on analogy to the tetracyclic system **66**.

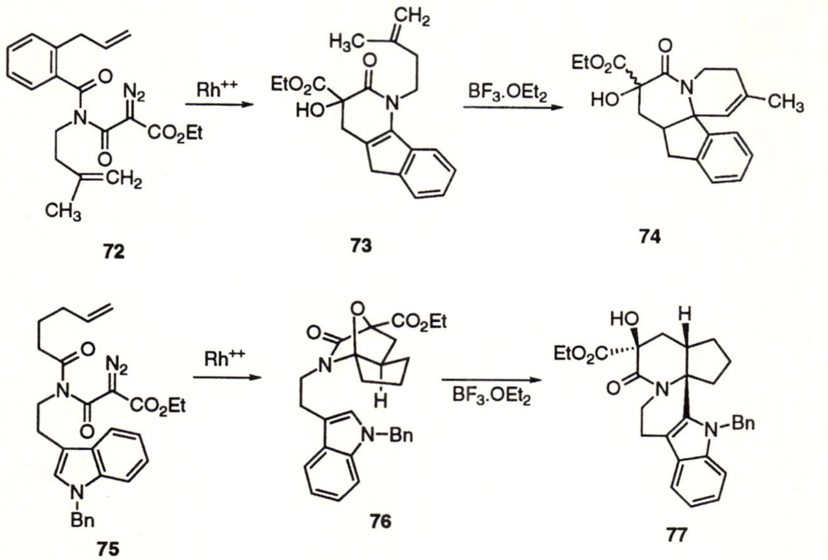

Scheme 18

We used this method as the key sequence in the synthesis of (±)-lycopodine (**78**). The intramolecular isomünchnone cycloadduct **81** was envisaged as the precursor of the key Stork intermediate **79** (via **80**) [42]. The heart of our synthetic plan was the formation of the latter intermediate by a Pictet-Spengler cyclization of the *N*-acyliminium ion derived from **81**. Central to this strategy was the expectation that the bicyclic iminium ion originating from **81** would exist in a chair-like conformation [42, 43]. Cyclization of the aromatic ring onto the iminium ion center should take place readily from the axial position. The readily available heptenoic acid **82** would serve as the precursor for the α-diazoimide, the direct progenitor of the isomünchnone dipole. This extension of the tandem cycloaddition-cationic 1-cyclization protocol to the formal synthesis of (±)-lycopodine (**78**) is outlined below.

Piperidine **79** was synthesized by the Barton-McCombie reaction [44] of **80** which gave the expected amido-ester (96%) as a 3:2-mixture of diastereomers. The mixture was hydrolyzed to the corresponding carboxylic acid which, upon thermal decarboxylation, gave the desired *N*-benzyl lactam (85% overall yield) as a single diastereomer. The structure was unequivocally established by a single-crystal X-ray analysis. Reduction of the lactam with $LiAlH_4$ (81%) followed by debenzylation via catalytic hydrogenation (Pd/C) afforded the key Stork intermediate **79** [42]. The preparation of **79** constitutes a total synthesis of (±)-lycopodine (**78**) and is based on a sequential dipolar-cycloaddition *N*-acyliminium ion cyclization. This approach is particularly attractive as the starting α-diazoimide can be prepared efficiently on a large scale and the cycloaddition and cyclization reactions are highly stereospecific. We are currently investigating the application of the methodology outlined here to other alkaloidal targets.

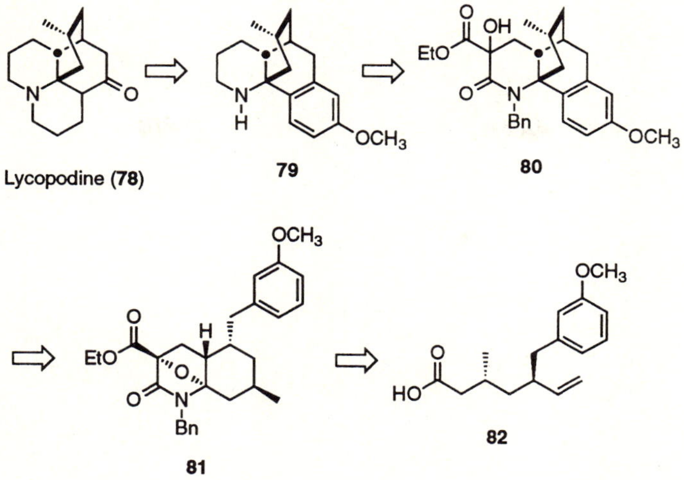

Lycopodine (78) 79 80

81 82

Scheme 19

5
Cycloadditions Across Heteroaromatic π-Systems

Given the propensity for isomünchnones to undergo dipolar cycloaddition with electron-rich dipolarophiles, systems in which the alkenyl group was incorporated into an electron-rich heteroaromatic ring were also studied [28]. Nitrile oxides and nitrile imines are known to undergo intramolecular 1,3-dipolar cycloaddition with furan and thiophenes [45–48]. This observation led our group to synthesize furanyl diazoimides **83** and **88** with the hope that intramolecular cycloaddition across the heteroaromatic system would occur. The Rh(II)-catalyzed reaction of **83**, however, failed to give the desired furanyl cycloadduct **84**. However, in the presence of DMAD a novel sequence of cycloadditions occurred. The initial transient isomünchnone **85** first underwent bimolecular cycloaddition with DMAD to provide cycloadduct **86** which, in turn, underwent a subsequent intramolecular Diels-Alder reaction to give polycycle **87** [28].

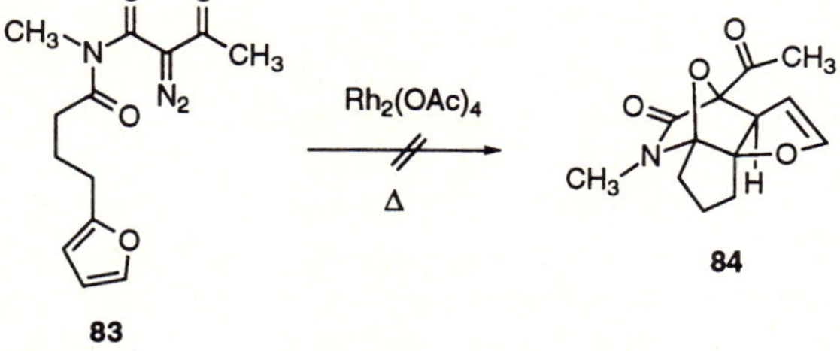

$Rh_2(OAc)_4$, Δ

83 84

Scheme 20

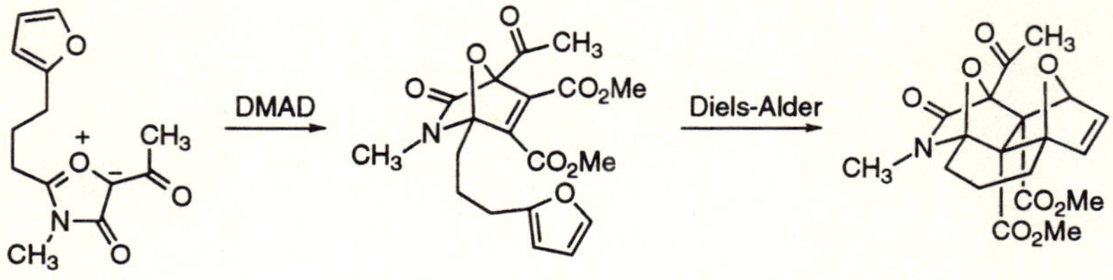

Scheme 21

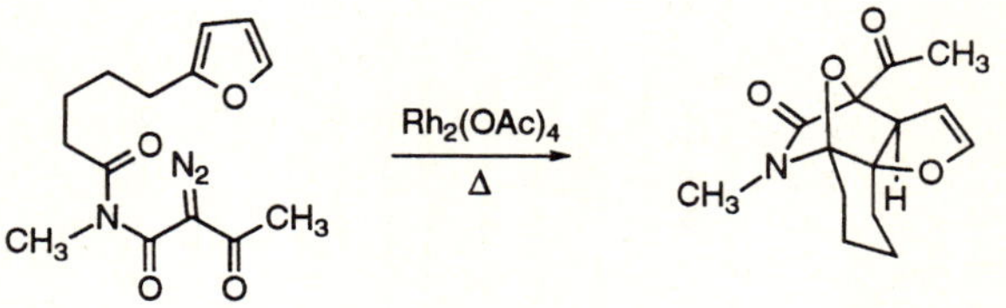

Scheme 22

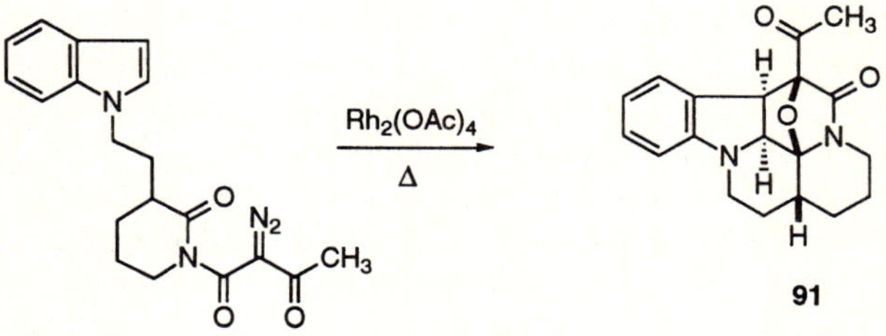

Scheme 23

As was mentioned earlier, the chain length of the tethered alkenyl group can influence the outcome of the cycloaddition reaction. When the chain length between the furanyl and isomünchnone ring was increased by one methylene unit, as in **88**, intramolecular dipolar-cycloaddition occurred producing cycloadduct **89** in high yield [49]. The ability of diazoimide **88** to undergo the intramolecular cycloaddition is presumably due to orbital overlap between the dipole and dipolarophile which is undoubtedly assisted by formation of the six-membered ring.

Our group has also encountered success in cycloadding an isomünchnone dipole across an indole double bond [28]. Cycloadduct **91** was generated in high yield as a single diastereomer from the $Rh_2(OAc)_4$-catalyzed reaction of diazoimide **90.** The assignment was unequivocally established by an X-ray crystal structure. The ready construction of these poly-heterocycles in one step and in high overall yield clearly demonstrates the potential of intramolecular dipolar-cycloadditions of isomünchnones as a strategy for natural product synthesis.

6
Cycloadditions Across Triple Bonds

The cycloaddition of isomünchnones with acetylenic dipolarophiles followed by the extrusion of an alkyl or aryl isocyanate (RNCO) has proven to be an effective method for the synthesis of substituted furans. The Ibata group investigated the bimolecular 1,3-dipolar-cycloaddition of aryl-substituted isomünchnones with a number of acetylenic dipolarophiles [50]. Aryl diazoimides of type **1** were heated in the presence of a catalytic amount of $Cu(acac)_2$ and the appropriate acetylenic dipolarophile. Formation of the substituted furan was found to be temperature-dependent; higher temperatures (ca. 120 °C) were needed for complete conversion to the furan. It was reasoned that the extrusion of methyl isocyanate was not as facile as the loss of carbon dioxide from sydnones and münchnones [50].

Non-aryl substituted isomünchnones also undergo the same transformation but under less rigorous conditions. Thus, when acyclic diazoimides **19** and **20** were subjected to $Rh_2(OAc)_4$-catalyzed decomposition in the presence of DMAD, cycloaddition followed by extrusion of methyl isocyanate occurred to give the substituted furans **94** and **95** [35].

Instead of losing methyl isocyanate, the extrusion of a tethered alkyl isocyanate occurred when the bicyclic diazoimide **96** was used. The rhodium(II) acetate-catalyzed reaction of **96** in the presence of DMAD produced furano-isocyanate **98** in 85% yield. The anticipated cycloadduct **97** was not isolated, but instead underwent a subsequent [4+2] cycloreversion under the reaction conditions to give the observed product. The initially formed furanoisocyanate **98** was

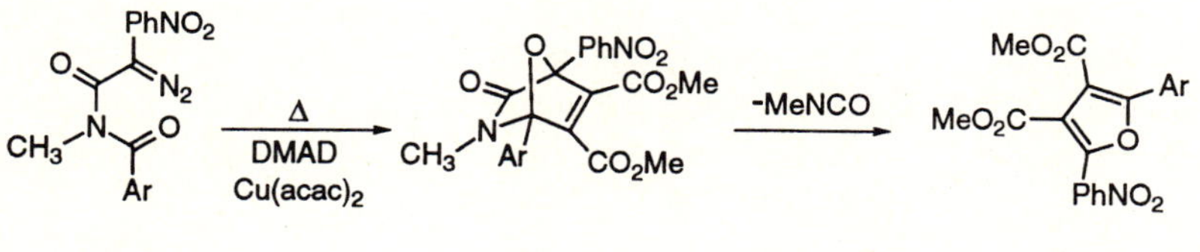

PhNO2
N2
CH3
Ar
Δ
DMAD
Cu(acac)2
PhNO2
CO2Me
CH3
Ar
CO2Me
-MeNCO
MeO2C
Ar
MeO2C
PhNO2

1 **92** **93**

Scheme 24

R
N2
CH3
CH3
Rh2(OAc)4
Δ
DMAD
-CH3NCO
MeO2C
MeO2C
CH3
R

19; R=CH3
20; R=CH2CH3

94; R=CH3
95; R=CH2CH3

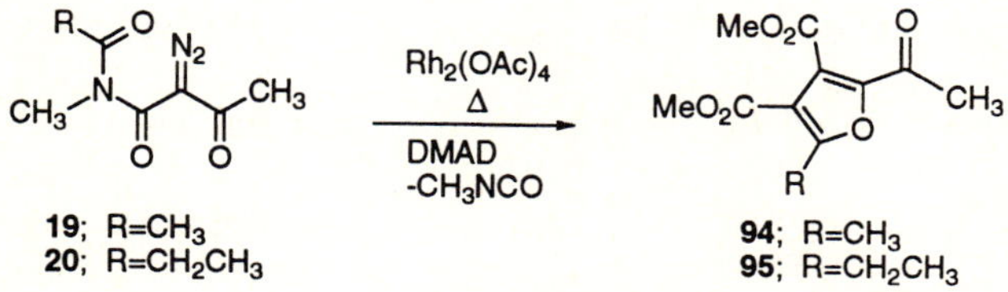

Scheme 25

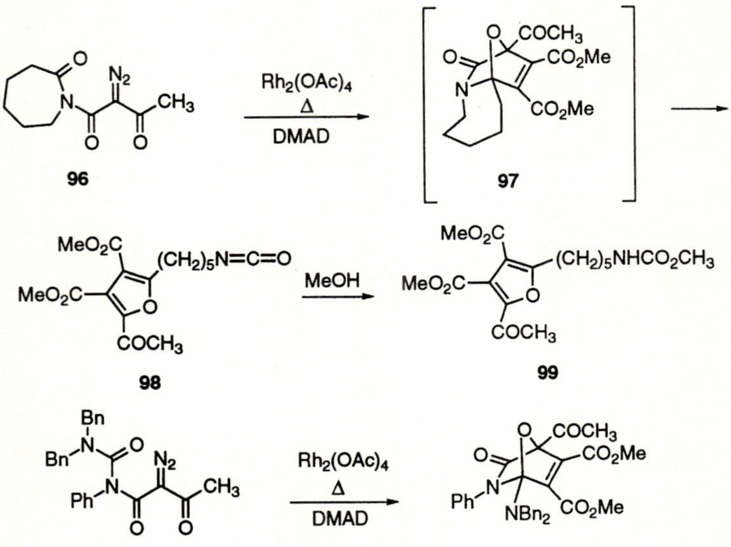

Scheme 26

characterized as its urethane derivative **99** by reaction with methanol [26–28]. Interestingly, treatment of the structurally related dibenzyl(diazoacetyl)urea **100** with $Rh_2(OAc)_4$ and DMAD afforded cycloadduct **101** which was stable enough to be isolated [26–28].

Several additional examples of the intramolecular cycloaddition of unactivated acetylenes with isomünchnones were reported by Maier [30]. This cycloaddition approach represents an efficient method for providing rapid access to annulated furans present in several sesqui- and diterpenes, such as the paniculides [51], furanonaphthoquinones [52], furodysin, and furodysinin [53, 54]. The decomposition of acyclic acetylenic diazoimides **102** and **103** with $Rh_2(OAc)_4$ resulted in cycloaddition and retro-Diels-Alder extrusion of methyl isocyanate to give annulated furans **104** and **105** in good yield. The overall transformation is closely related to the intramolecular Diels-Alder reactions of acetylenic oxazoles extensively studied by Jacobi and coworkers [55].

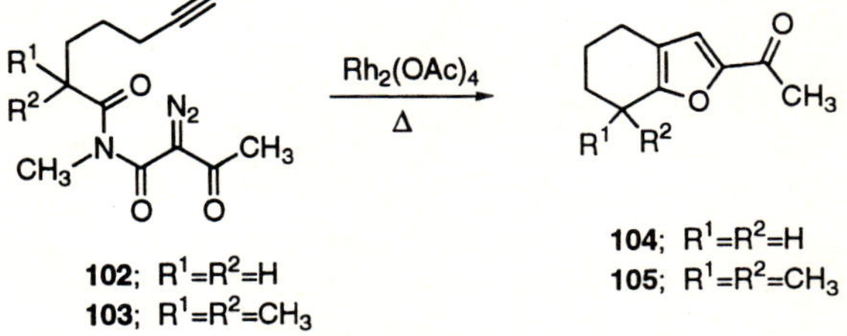

Scheme 27

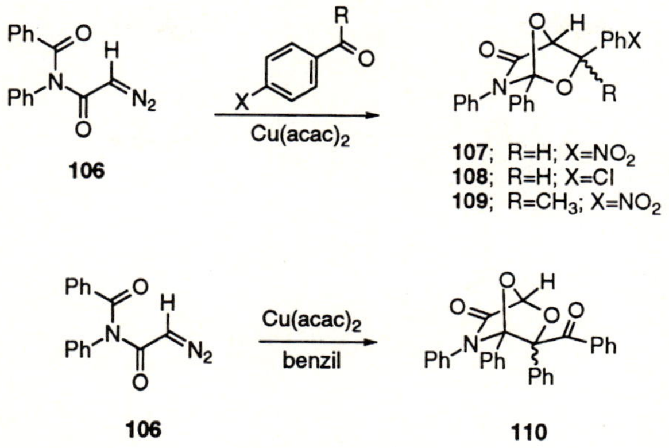

Scheme 28

An interesting feature of isomünchnones is their ability to undergo 1,3-dipolar cycloaddition with carbonyl compounds, a reaction which is unprecedented with münchnones [56]. This is illustrated by the reaction of diazoimide **106** with Cu(acac)$_2$ in the presence of several different aldehydes and ketones which resulted in the formation of cycloadducts of type **107**–**109**. When benzil was used as the dipolarophile, the regioselectivity was reversed giving rise to cycloadduct **110** as the only regioisomer.

7 Formation of Azomethine Ylides Derived from Imines and Oximes

The interaction of a metallo carbene with an imine nitrogen atom to give a transient azomethine ylide has attracted attention over the past decade [57]. Some of the standard methods for generating azomethine ylides involve the thermal or photolytic ring opening of aziridines [58], desilylation [59], or dehydrohalogenation [60] of iminium salts, and proton abstraction from imine derivatives of α-amino acids [61]. Azomethine ylides are of interest because these dipoles undergo facile 1,3-dipolar cycloaddition with π-bonds to give pyrrolidines which, in turn, have been used to prepare a variety of alkaloids [62].

The tandem reaction of carbenoids with simple imines to form azomethine ylides which then undergo 1,3-dipolar cycloaddition with various dipolarophiles was first reported in 1972 [63]. Thus, treatment of phenyldiazomethane with copper bronze in the presence of excess *N*-benzylidene-methylamine resulted in the isolation of imidazoline **112**. Formation of this product was rationalized by carbenoid addition onto the imine nitrogen to give azomethine ylide **111** which then underwent a 1,3-dipolar cycloaddition with another molecule of imine to produce the observed product. Bartnik and Mloston subsequently extended this observation by using other dipolarophiles [64]. For example, catalytic decomposition of phenyl-diazomethane and *N*-benzylidene-methylamine in the presence of dimethyl maleate or benzaldehyde gave pyrrolidine **113** and oxazolidine **114**, respectively. In both cases, no product resulting from the trapping of the ylide

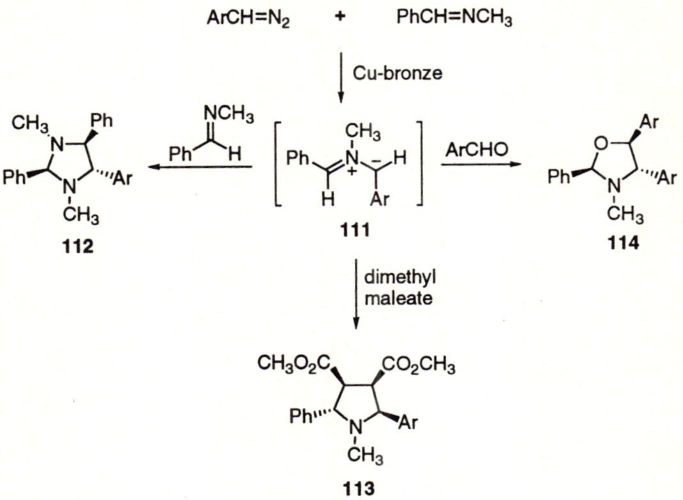

Scheme 29

with a molecule of imine could be observed. Catalytic decomposition of phenyldiazomethane with other Schiff bases was found to proceed via formation of a *trans*-1,3-dipole. Depending on the size and quantity of the substituent groups, the ylide either undergoes cyclization in a conrotatory sense to a *cis*-aziridine or [3+2]-cycloaddition to an available π-bond. The reactivity of double bonds toward the ylide was found to decrease in the order C=C → C=O → C=N.

Since they were first isolated from penicillins, thiazoloazetidinones such as **115** have become versatile intermediates in the synthesis of various β-lactam antibiotics. Soft electrophiles prefer to attack at the sulfur atom whereas hard electrophiles react with the thiazoline nitrogen. Thomas and coworkers have investigated the reaction of thiazoloazetidinone **115** with metal carbenoids [65]. Treatment of **115** with a large excess of ethyl diazoacetate in the presence of copper (II) acetoacetonate and dimethyl fumarate gave the bis-methoxycarbonyl adduct **117**. The formation of this material involves an initial addition of the ethoxycarbonyl carbenoid onto the thiazoline nitrogen to produce azomethine ylide **116**. This reactive dipole undergoes a subsequent 1,3-dipolar cycloaddition with the added dipolarophile to give the observed product. The reaction was found to be both regio- and stereoselective. No products derived from the reaction of the carbenoid at the sulfur atom or at the C–C double bond were observed. The stereochemistry at C-3 of the cycloadduct is consistent with approach of the fumarate ester from the less hindered side of the ylide.

The formation and intramolecular dipolar cycloaddition of azomethine ylides formed by carbenoid reaction with C–N double bonds has recently been studied by the author's group [66]. Treatment of 2-(diazoacetyl)benzaldehyde *O*-methyl oxime (**118**) with rhodium (II) octanoate in the presence of dimethyl acetylenedicarboxylate or *N*-phenylmaleimide produced cycloadducts **120** and

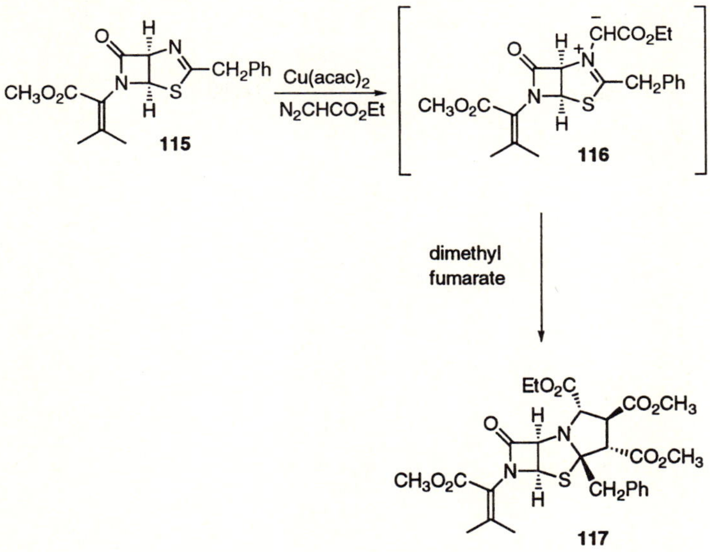

Scheme 30

121, respectively. The cycloaddition was also carried out using *p*-quinone as the dipolarophile. The major product isolated corresponded to cycloadduct **122.** The subsequent reaction of this material with excess acetic anhydride in pyridine afforded diacetate **123** in 67% overall yield from **118.** This latter compound incorporates the basic dibenzo[a,d]cyclohepten-5,10-imine skeleton found in MK-801 [67], which is a selective ligand for brain cyclidine (PCP) receptors that has attracted considerable attention as a potent anticonvulsive and neuro-protective agent [68, 69].

The oxime nitrogen lone pair of electrons must be properly oriented so as to interact with the rhodium carbenoid [66]. Thus, subjection of the *E*-oximino isomer **124** to a catalytic quantity of $Rh_2(OAc)_4$ in CH_2Cl_2 (40 °C) with a slight excess of DMAD afforded the bimolecular cycloadduct **126** in 93% yield. In sharp contrast, when the isomeric *Z*-oximino diazo derivative **125** was exposed to the same reaction conditions, only indanone-oxime **127** (80%) was obtained. The formation of this product probably occurs by an intramolecular C-H insertion reaction.

The success achieved with the Rh(II)-catalyzed transformations of *E*-oximino diazo carbonyl compounds prompted our group to study some additional systems where the C-N π-bond was configurationally locked so that azomethine ylide formation would readily occur. To this end, we investigated the Rh(II)-catalyzed behavior of isoxazoline **128** in the presence of DMAD. This reaction afforded the azomethine-derived cycloadduct **129** as a 4:1-mixture of diastereomers in 65% yield. A similar transformation occurred using the α-diazoacetophenone derivative **130** which produced isoxazolo[3, 2-a]isoquinoline **131** as a 2:1-mixture of diastereomers in 82% yield.

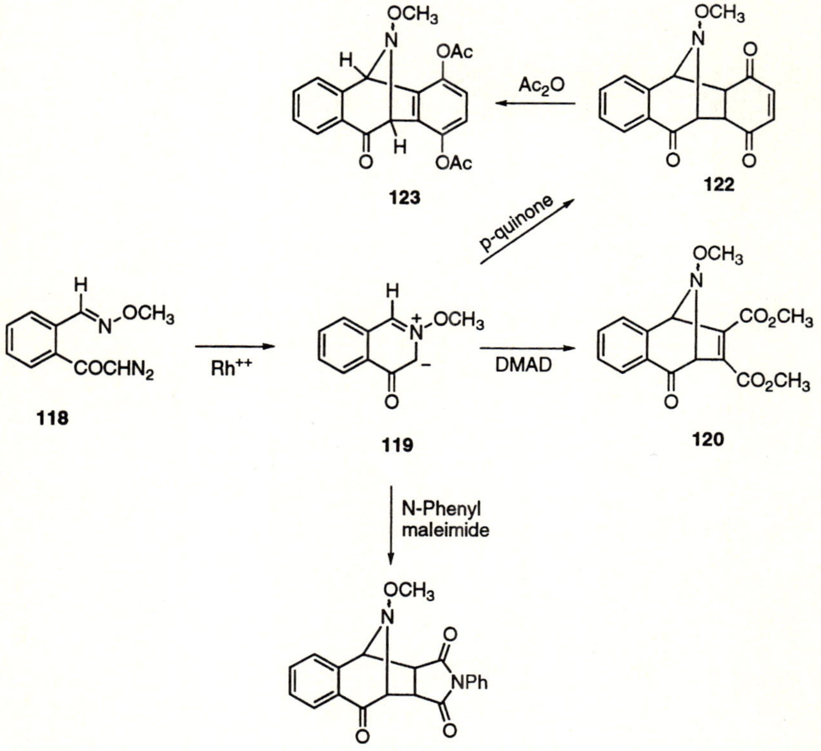

OCH3 N H OAc Ac2O OCH3 N O O O H OAc O O 123 122 p-quinone OCH3 N H OCH3 H N+ OCH3 CO2CH3 N COCHN2 Rh++ – DMAD CO2CH3 O O O 118 119 120 N-Phenyl maleimide OCH3 N O NPh O O 121

Scheme 31

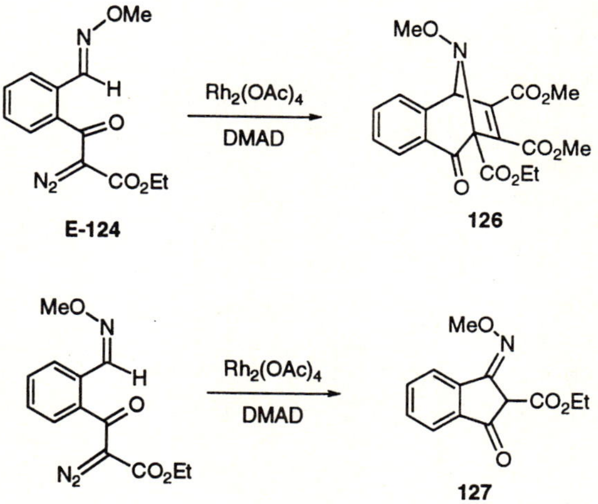

OMe N MeO N H Rh2(OAc)4 CO2Me O DMAD CO2Me N2 CO2Et O CO2Et E-124 126

MeO N MeO N H Rh2(OAc)4 O DMAD CO2Et O N2 CO2Et Z-125 127

Scheme 32

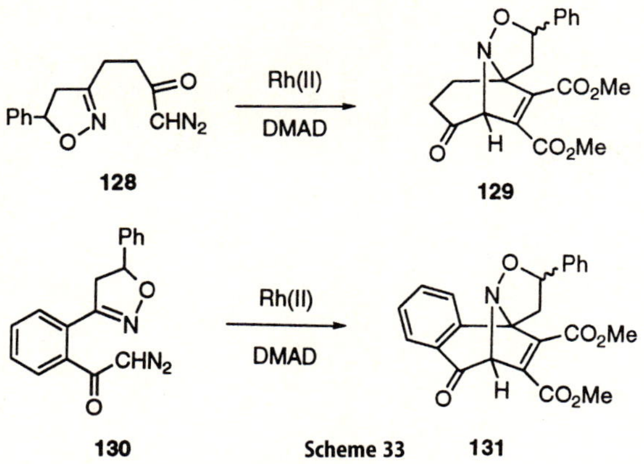

Scheme 33

8 Pyridinium Ylides

Since their introduction in 1960 [70], pyridine ylides have become increasingly popular probes of the dynamics of carbenes which lack chromophores [71–76]. The combination of high reactivity, favorable spectroscopic properties, and long ylide lifetime has allowed the study of the dynamics of a variety of "invisible" carbenes [77]. The technique has found use in the study of aryl, arylhalo, alkyl, alkylalkoxy, alkylhalo, arylsiloxy, and dialkyl carbenes [78–81]. A number of examples dealing with the preparation of stable pyridinium ylides have also been reported in the literature [82–85]. Pyridinium tetraphenylcyclopentadienylide (**133**) was synthesized by irradiating 2,3,4,5-tetraphenyl-diazocyclopentadiene (**132**) in pyridine. Addition of water precipitated the purple ylide **133** in almost quantitative yield [82–84]. This process appears to be general for a number of substituted pyridines (i.e. 2-picoline, 3-picoline, and 2,6-lutidine). In an analogous fashion, *N*-dicyanomethylide **134** was prepared from the photolysis of diazomethane-dicarbonitrile in pyridine [85].

Although the transition metal-catalyzed reaction of α-diazocarbonyl compounds with aromatic molecules has received much attention in recent years [86], the metal-catalyzed behavior of these compounds with N-containing

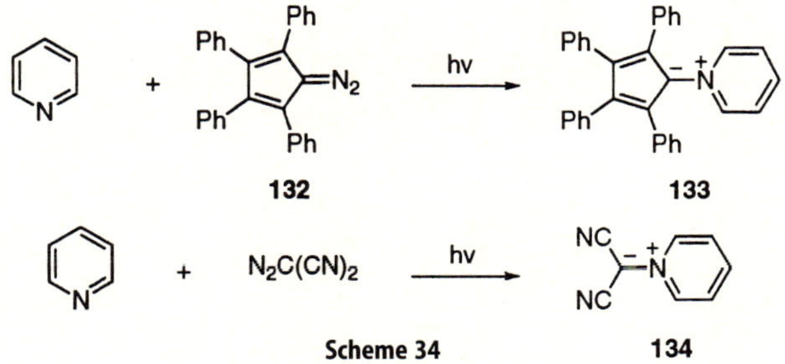

Scheme 34

heteroaromatics has not been extensively studied. An early example involved the preparation of isoquinoline-carboethoxymethylide **135** by the thermal decomposition of ethyl diazoacetate in the presence of isoquinoline [87]. The same ylide could also be obtained from *N*-carboethoxymethylene isoquinolinium bromide by the elimination of hydrogen bromide. Ylide **135** is a red crystalline solid which is stable in the absence of moisture. The dipolar character of **135** was established by its reaction with dimethyl acetylenedicarboxylate which led to the formation of cycloadduct **136.** Platz and coworkers reported that the photolysis of phenylchlorodiazirine **137** in the presence of both pyridine and DMAD produced cycloadduct **139** in 30% yield by dipolar-cycloaddition of DMAD to the ylide followed by loss of HCl [88].

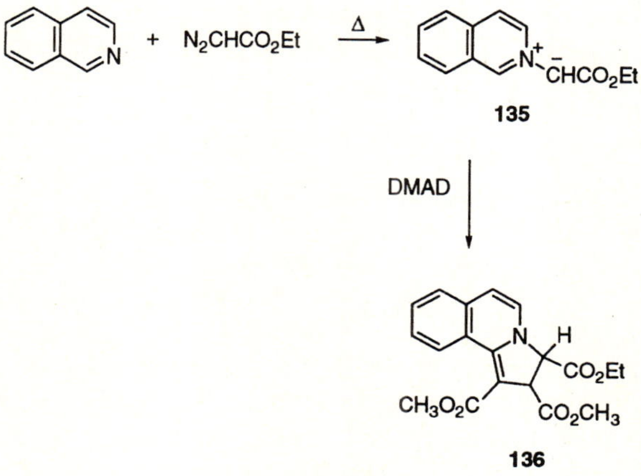

+ N_2CHCO_2Et Δ N^+ $\bar{C}HCO_2Et$ **135** DMAD N H CO_2Et CH_3O_2C CO_2CH_3 **136**

Scheme 35

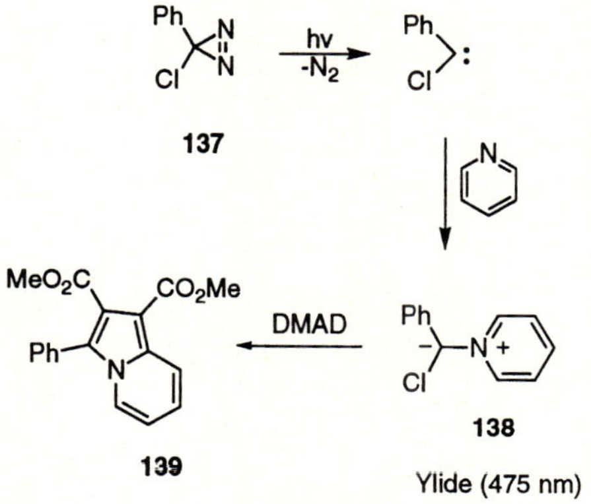

Ph N N Cl **137** hν $-N_2$ Ph Cl : N Ph Cl N+ **138** Ylide (475 nm) DMAD MeO_2C CO_2Me Ph N **139**

Scheme 36

As part of our group's continuing involvement with the chemistry of azomethine ylides, we became interested in examining the cyclization of α-diazo substituted N-containing heteroaromatic systems as a method for ylide generation. Apart from the above examples using pyridines [78] and isoquinolines [87], little was known about the diazo cyclization process with *N*-heteroaromatic systems when we initiated our work in this area [89]. The Rh(II)-catalyzed reaction of α-diazoacetophenone in the presence of 2-methylthio-pyridine and dimethyl acetylenedicarboxylate gave 3-benzoyl-1,2-dicarbomethoxy-3,5-dihydro-5-methylthioindolizine (**143**). The formation of **143** proceeds via a pyridinium ylide formed by attack of the nitrogen lone pair on the electrophilic keto carbenoid. Subsequent dipolar cycloaddition of ylide **141** with DMAD occurs at the less substituted carbon atom to give cycloadduct **142**. This transient species is converted to **143** by means of a 1,5-sigmatropic hydrogen shift. The results are also consistent with the formation of the regio-isomeric cycloadduct **144** which undergoes a 1,5-thiomethyl shift, perhaps via the tight ion pair **145**.

A related cyclization occurred using 1-diazo-3-[(2-(pyridyl)thio]-2-propanone (**146**). The initial reaction involves generation of the expected pyridinium ion **147** by intramolecular cyclization of the keto carbenoid onto the nitrogen atom of the pyridine ring. Dipolar cycloaddition of **147** with DMAD affords

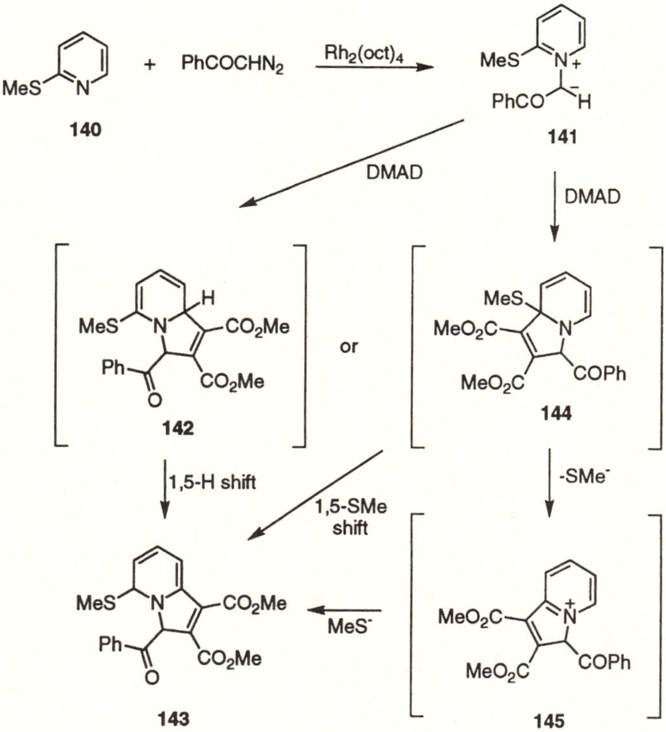

Scheme 37

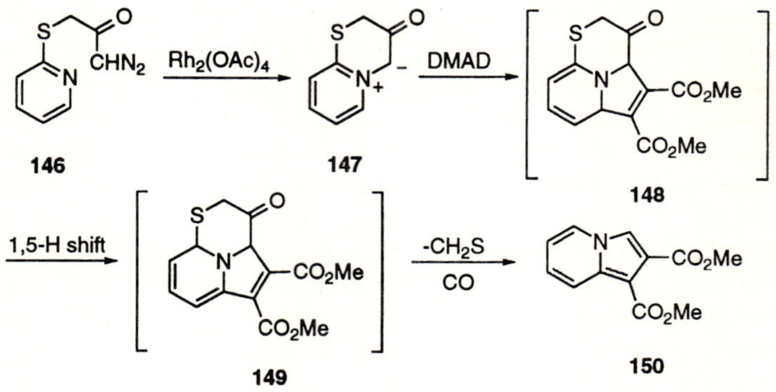

Scheme 38

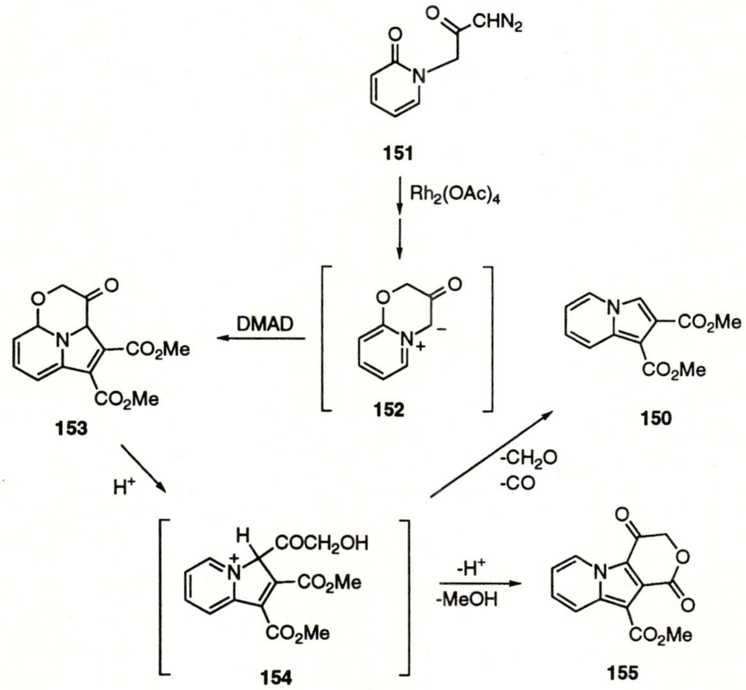

Scheme 39

cycloadduct **148** which undergoes a subsequent 1,5-hydrogen shift to give **149** followed by fragmentation of CO and CH_2S to produce indolizine **150.**

Interestingly, the Rh(II)-catalyzed reaction of 1-(3′-diazo)-acetonyl-2-pyridone (**151**) with DMAD was found to give cycloadducts derived from an azomethine ylide. The initial reaction involves generation of the expected carbonyl ylide dipole by intramolecular cyclization of the keto carbenoid onto the oxygen atom of the amide group. A subsequent proton exchange generates the thermodynamically more stable azomethine ylide **152** which is trapped by DMAD, eventually producing cycloadduct **153**. The formation of products **150** and **155** from cycloadduct **153** proceeds by an acid-catalyzed C–O bond cleavage giving pyridinium ion **154**. This transient species can lose a proton and lactonize to **155** or else undergo fragmentation to afford formaldehyde, carbon monoxide and indolizine **150**.

Azomethine ylide cycloadducts derived from keto carbenoid cyclization onto a thiobenzoxazole have also been encountered in our studies. When 1-diazo-3-[2-benzoxazolyl)thio]-2-propanone (**156**) was used, the initially formed cycloadduct **158** undergoes a subsequent 1,3-sigmatropic thio shift to give the thermodynamically more stable product **159**. A good analogy can be found in the literature for the suggested 1,3-sigmatropic shift [90].

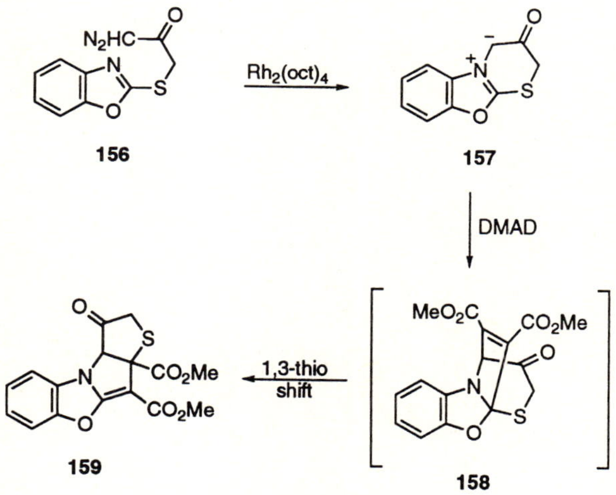

Scheme 40

An entirely different reaction occurred when 2-(4-diazo-3-oxobutyl)-benzoxazole (**160**) was treated with Rh(II) octanoate. In addition to undergoing dipolar cycloaddition to produce cycloadduct **162** (20%), the highly stabilized dipole (i.e. **161**) formed from the benzoxazole isomerized by proton exchange to produce the cyclic ketene N,O-acetal **163**. This compound reacted further with the activated 1-bond of DMAD to give zwitterion **164**. The anionic portion of **164** then added to the adjacent carbonyl group, producing a new zwitterionic intermediate **165**. In the presence of water, this species was converted to the phenolic lactam **166**.

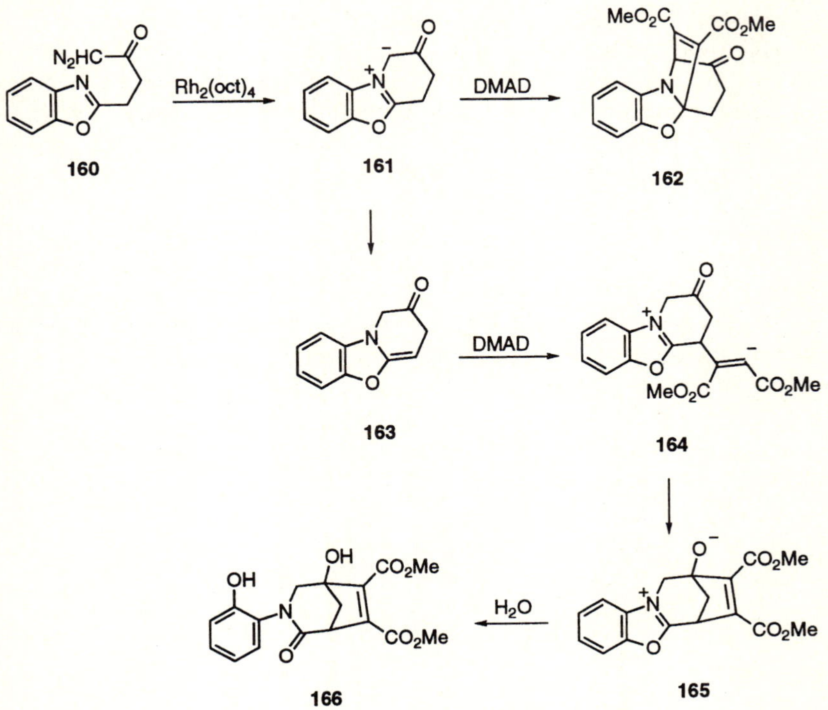

Scheme 41

9
Nitrogen Ylides Derived From Diazocarbonyls and Nitriles

1,3-Oxazoles with various substitution patterns are well-known heterocycles for which a number of methods of synthesis have been reported [91]. Acyl carbenes or functionally equivalent species have been found to undergo cyclization with nitriles to give oxazoles in high yield via nitrile ylide intermediates [92, 93]. This reaction can be induced to occur under thermal, photolytic or catalytic conditions [91, 94, 95]. Huisgen and coworkers were the first to study this process in some detail [94]. Thermolysis (or copper catalysis) of a mixture of ethyl diazoacetate and benzonitrile resulted in the formation of oxazole **168**. The isolation of this product is most consistent with a mechanism involving metallo carbene addition onto the nitrile nitrogen atom to generate dipole **167** which then cyclizes to produce oxazole **168**.

Dimethyl diazomalonate undergoes reaction with nitriles in the presence of rhodium (II) acetate to give 2-substituted-4-carbomethoxy-1,3-oxazoles (**169**). The reaction proceeds with a wide range of nitriles [95–101]; however, cyclopropanation is a competing process in the case of unsaturated nitriles [91].

Kende and coworkers have reported on the formation of a nitrile ylide intermediate from carbenes and methyl acrylonitrile. Thermolysis of *p*-diazooxide

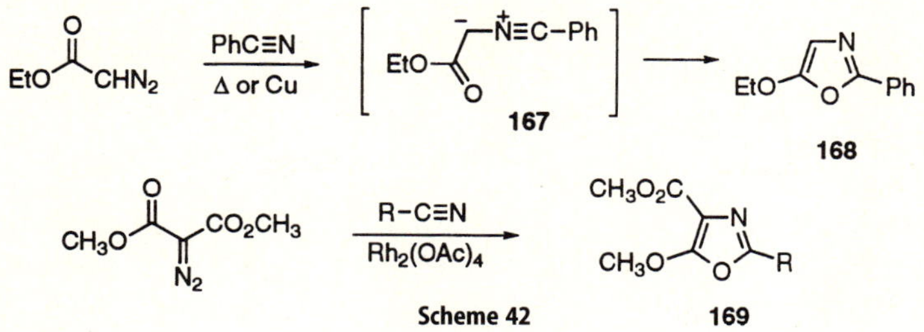

Scheme 42

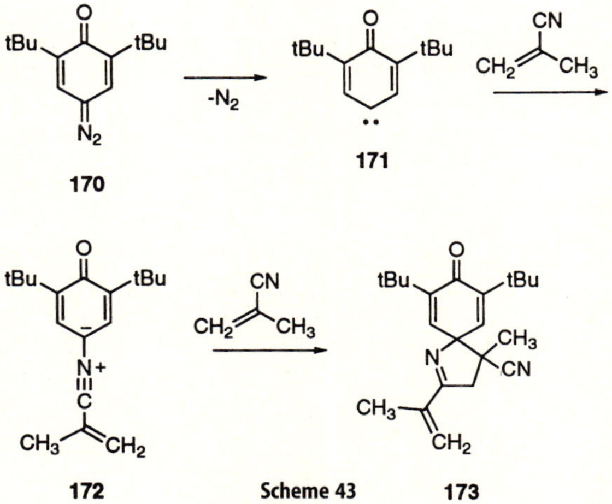

Scheme 43

170 in methyl acrylonitrile as solvent gave the spirocyclic product **173** in 48% yield [102].The formation of **173** was interpreted in terms of the generation of nitrile ylide **172** followed by 1,3-dipolar cycloaddition across the C–C double bond of a second molecule of methylacrylonitrile. The regiochemistry of the cycloaddition is consistent with FMO theory.

In a somewhat similar manner, diazodicyanoimidazole (**174**) was found to give the fused heterocycle **176** when heated in benzonitrile [103].This reaction presumably involves the intermediacy of nitrile ylide **175**.

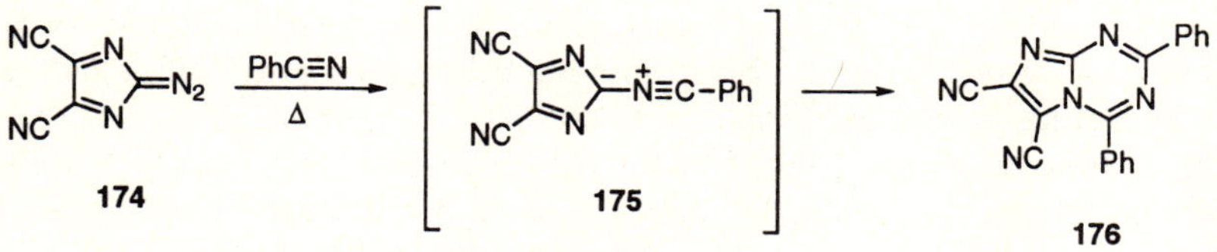

Scheme 44

10 Dipole Cascade Processes

1,3-Dipoles are extremely valuable intermediates in synthetic organic chemistry. Their best known reaction corresponds to a 1,3-dipolar-cycloaddition reaction [62]. Less attention, however, has been attached to the interconversion of one dipole to another [104–108]. Rearrangement of 1,3-dipoles is encountered far less frequently than analogous carbocation [109–111], carbene [112, 113], or radical reorganizations [114–116]. Those rearrangements which do occur can be divided into a small number of types, defined either by the overall structural change or by the nature of the individual steps involved. Several years ago our research group introduced a new method for azomethine ylide formation in which the key step involved a dipole rearrangement. This reaction, which we have termed a "dipole cascade" involves three distinct classes of 1,3-dipoles [117]. It is initiated by a rhodium(II)-catalyzed α-diazo ketone (**177**) cyclization onto a neighboring carbonyl group to generate a carbonyl ylide dipole (**178**) which then undergoes a subsequent proton shift to give an azomethine ylide (**179**).

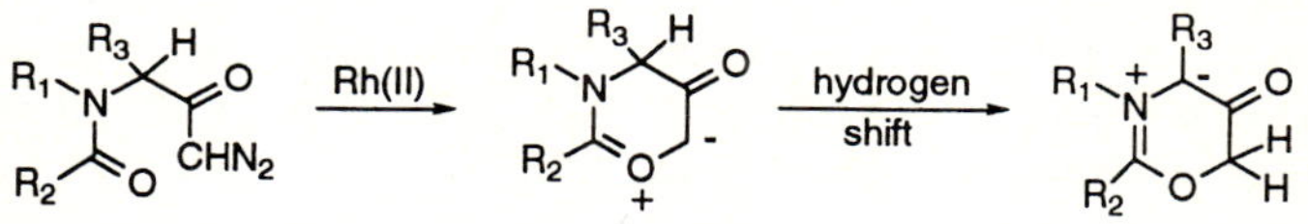

Rh(II)

hydrogen shift

177 **178** **179**

Scheme 45

The wealth of strategically located functionalities that result from this novel cascade process was uncovered during an examination of the reaction of (*S*)-1-acetyl-2-(1-diazoacetyl)pyrrolidine (**180**) with 1.5 equiv of dimethyl acetylenedicarboxylate in the presence of a catalytic quantity of rhodium(II) acetate. Very little (<10%) of the expected carbonyl ylide-derived cycloadduct (i.e. **182**) was obtained [117]. Instead, the major product (90%) corresponded to structure **185**. A mechanism that rationalizes the formation of this product involves generation of the expected carbonyl ylide dipole **181** by intramolecular cyclization of the keto carbenoid onto the oxygen atom of the amide group. Isomerization of **181** to the thermodynamically more stable azomethine ylide **183** occurs via proton exchange with a small amount of water that was present in the reaction mixture. 1,3-Dipolar cycloaddition with dimethyl acetylenedicarboxylate provides cycloadduct **184**, which undergoes a subsequent 1,3-alkoxy shift to generate the tricyclic dihydropyrrolizine **185**. MNDO calculations show that cyclic carbonyl ylides of type **181** have higher heats of formation (ca. 15 kcal/mol) than the corresponding azomethine ylide **183**. Some of this energy difference is presumably responsible for the ease with which the dipole reorganization occurs.

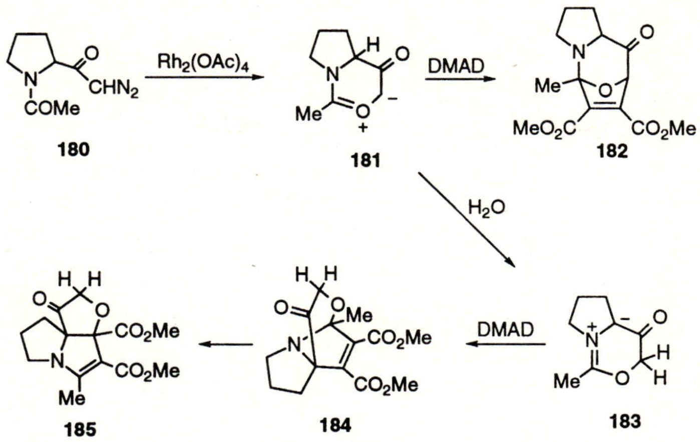

Scheme 46

In the dipole cascade reaction, a proton must be removed from the α-carbon atom in order to generate the azomethine ylide. When the α-position of the pyrrolidine ring was blocked by a benzyl group, formation of the azomethine ylide dipole could not occur. In fact, treatment of diazoketone **186** with rhodium(II) acetate in the presence of dimethyl acetylenedicarboxylate afforded only the carbonyl ylide-derived cycloadduct **187** in 95% yield [117].

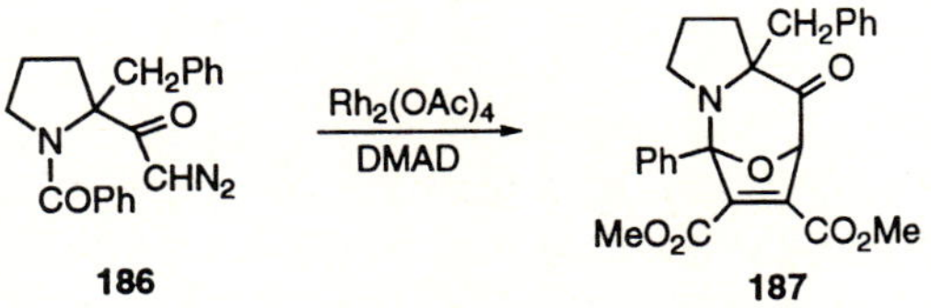

Scheme 47

A further example of the dipole cascade process was encountered in a study of the Rh(II)-catalyzed decomposition of α-diazoketone **188** which gave the novel carbonyl rearrangement product **192** [118]. Intramolecular trapping of the rhodium carbenoid by the benzimidazolone carbonyl group generates the stabilized carbonyl ylide **189**. Collapse of **189** to the epoxide **190** followed by ring opening gave the zwitterion **191**. Attack of the alkoxide ion on the more electrophilic carbonyl (ketone vs ester) and carbon migration then gave product **192**.

When the reaction of **188** was carried out in the presence of DMAD, two unusual addition/rearrangement products were obtained and identified as compounds **196** and **199** in 33% and 24% yield, respectively. Under these conditions the rearrangement product **192** was not observed. Formation of the unexpected products **196** and **199** resulted from the trapping of two isomeric ylides. Bimole-

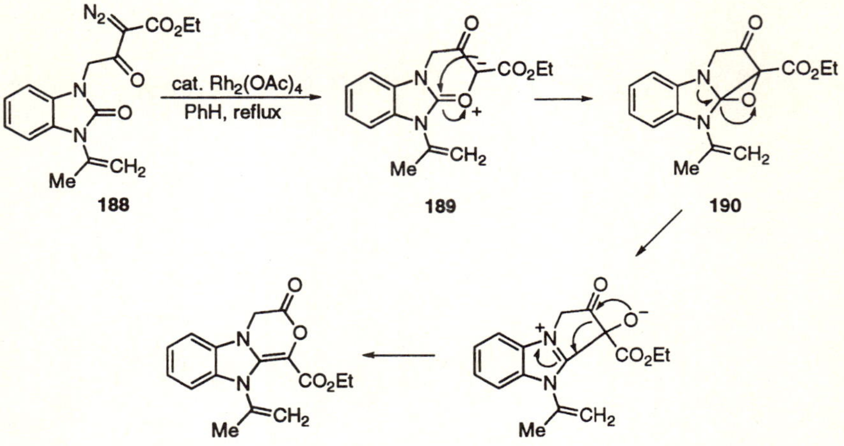

Scheme 48

cular cycloaddition of the expected carbonyl ylide **189** with DMAD gave the [3+2]-cycloadduct **194**, which under the reaction conditions fragmented to zwitterion **195**. Carbon to oxygen acyl migration then generated the eight-membered dienol lactone **196**. Rearrangement of **195** to **196** can be seen as a vinylogous rearrangement of **191** to **192** and underscores the thermodynamic driving force for this type of transformation. Formation of **199** requires the tandem cascade of carbonyl ylide **189** to azomethine ylide **193**. Condensation of **193** with DMAD resulted in the [3+2]-cycloaddition product **197**. Fragmentation to zwitterion **198** followed by proton transfer eventually afforded pyrrolobenzimidazole **199**.

In the case of α-diazo ketoamide **200**, the carbonyl ylide dipole is sufficiently stabilized via resonance to be trapped by dimethyl acetylenedicarboxylate to give cycloadduct **201** in 90% yield [119]. No material derived from azomethine ylide cycloaddition was observed. The closely related α-diazo ketoamide **202** was also examined. Most interestingly, treatment of **202** with rhodium(II) acetate in the presence of dimethyl acetylenedicarboxylate afforded cycloadduct **203** in 60% yield. The initial reaction involved generation of the expected carbonyl ylide dipole **205** by intramolecular cyclization of the keto carbenoid onto the oxygen atom of the amide group. This highly stabilized dipole does not readily undergo 1,3-dipolar cycloaddition but rather isomerizes to the cyclic ketene N,O-acetal **206** by proton exchange. This material reacted further with the activated π-bond of the dipolarophile to produce zwitterion **207**. The anionic portion of **207** added to the adjacent carbonyl group, affording a new zwitterionic intermediate **208**. Under anhydrous conditions, epoxide formation occurred with charge dissipation to give the observed cycloadduct **209**. The high efficiency of the dipole cascade, in conjunction with the intriguing chemistry of the resulting cycloadducts, presents numerous synthetic possibilities for the preparation of complex heterocycles.

Scheme 49

Scheme 50

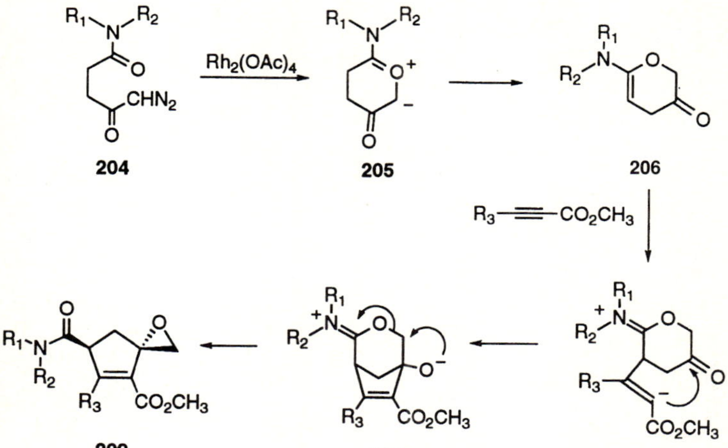

Scheme 51

11
Application of the Tandem Cyclization-Cycloaddition Sequence to the Pentacyclic Skeleton of the Aspidosperma Ring System

As mentioned above, our group has extensively employed the tandem cyclization-cycloaddition reaction of rhodium carbenoids as the key strategic element for the efficient syntheses of a wide variety of polycyclic nitrogen heterocycles. More recently, we have developed a fundamentally new approach to the construction of the pentacyclic skeleton of the aspidosperma ring system which involves a related domino cascade reaction [120]. This new strategy was successfully applied to the synthesis of desacetoxy-4-oxo-6,7-dihydrovindorosine (**211**).The approach used is shown below in antithetic format and is centered on the construction of the key oxabicyclic intermediate **212**. We reasoned that **211** should be accessible by reduction of **212**, which, by analogy with our previous work, should be available by the tandem rhodium(II)-catalyzed cyclization-cycloaddition of α-diazoimide **213**. Cycloaddition of the initially formed dipole across the pendant indole 1-system [49] would be expected to result in the simultaneous generation of the CD-rings of the aspidosperma skeleton [121]. The stereospecific nature of the internal cycloaddition reaction should also lead to the correct relative stereochemistry of the 4 chiral centers about the C-ring. In a recent publication, we described our initial experiments which verified the underlying viability of this approach to the aspidosperma skeleton [120].

The synthesis of α-diazoimide **213** commenced with the easily available β-ketoester **215**. *N*-Acylation of **215** with *N*-methylindole-3-acetyl chloride (**214**) using 4 Å molecular sieves as a neutral acid scavenger gave the desired imide (65%) which was readily converted to the requisite α-diazoimide **213** (90%) using standard diazo transfer methodology [122]. When α-diazoimide **213** was treated with a catalytic quantity of $Rh_2(OAc)_4$ in benzene at 50 °C, cycloadduct

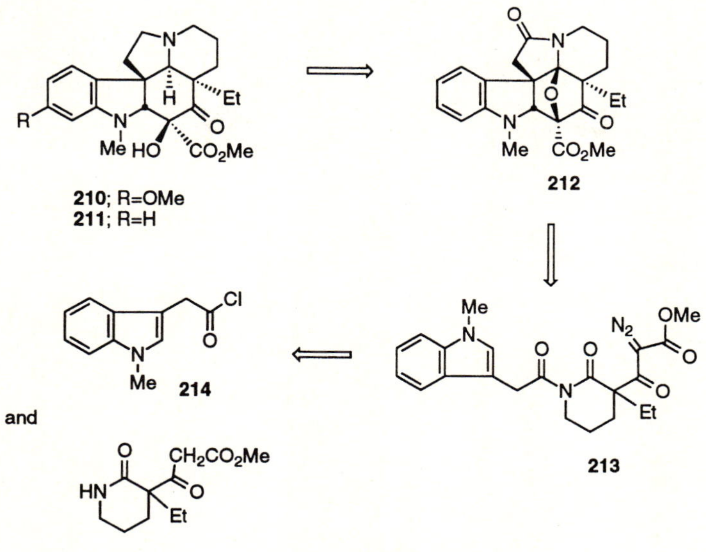

Scheme 52

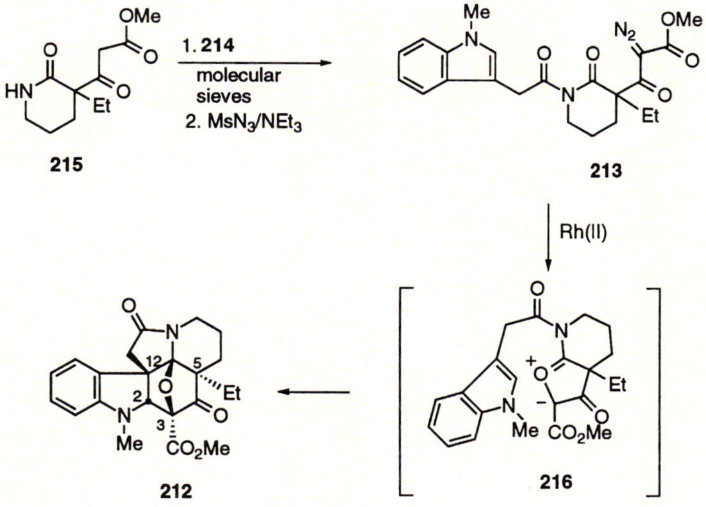

Scheme 53

212 was isolated in 95% yield as a single diastereomer. The structure of **212** was firmly established by X-ray crystallographic analysis which revealed that the cycloadduct contains the same relative stereochemical centers (C_2, C_3, C_5 and C_{12}) as those found in vindoline [123]. The formation of **212** occurs by cyclization of the initially formed rhodium carbenoid (derived from **213**) onto the neighboring piperidone carbonyl oxygen to give dipole **216** which subsequently cycloadds across the indole 1-bond. The isolation of **212** is the consequence of *endo* cycloaddition with regard to the dipole, and this is in full agreement with the lowest energy transition state. The cycloaddition can also be considered doubly diastereoselective in that the indole moiety approaches the dipole exclusively from the side of the ethyl group, away from the more sterically encumbered piperidone ring.

Cycloadduct **212** was subsequently converted to desacetoxy-4-oxo-6,7-dihydrovindorosine **211** via intermediate **217** in high overall yield thereby proving the merits of the method. The tandem cyclization-cycloaddition sequence is particularly attractive as four of the stereocenters are formed in one step with a high degree of stereocontrol.

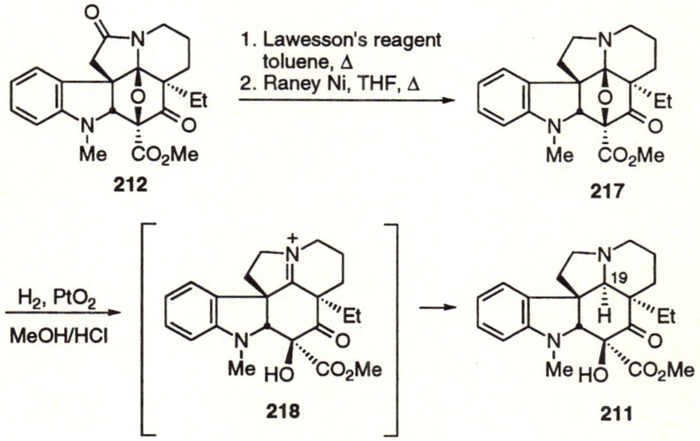

Scheme 54

12 Conclusion

Tandem ylide generation from the reaction of metallo carbenoids with nitrogen-containing substrates continues to be of great interest both mechanistically and synthetically. Effective ylide formation in transition metal-catalyzed reactions of α-diazo compounds depends on the catalyst, the α-diazo species, the nature of the substrate, and competition with other processes. The many structurally diverse and highly successful examples cited in this review clearly indicate that the tandem reaction of metallo carbenoids has evolved as an important strategy for the synthesis of polyaza heterocycles. It is reasonable to expect that future years will see the continued evolution of the cascade chemistry of transition metal carbenoids derived from α-diazocarbonyls in organic synthesis. As is the case in all new areas of research using catalysts, investigation of the chemistry of these transition metal complexes in the future will be dominated by the search for asymmetric synthesis.

Acknowledgment. I am indebted to my excellent coworkers for their commitment and considerable contributions in the form of ideas and experiments. Their names can be found in the references. The National Institutes of Health and the National Science Foundation are gratefully acknowledged for generous support of our research program.

References

1. Ho TL (1992) Tandem organic reactions. Wiley, New York
2. Ziegler FE (1991) Combining C-C π-bonds. In: Paquette LA (ed) Comprehensive organic synthesis. Pergamon, Oxford, vol 5, chap 7.3
3. Tietze LF, Beifuss U (1993) Angew Chem Int Ed Engl 32:131.
4. Waldmann H (1995) Domino reaction in organic synthesis highlight II. VCH, Weinheim, pp 193–202
5. Curran DP (1991) In: Trost BM, Fleming I (eds) Comprehensive organic synthesis. Pergamon, Oxford, vol 4, p 779
6. Wender PA (ed) (1996) Frontiers in organic synthesis. Chem Rev 96, pp 1–600
7. Bartlett PA (1984) In: Morrison JD (ed) Asymmetric synthesis. Academic Press, Orlando, vol 3. Fish PV, Johnson WS (1994) J Org Chem 59:2324. Jacobsen EJ, Levin J, Overman LE (1988) J Am Chem Soc 110:4329. Johnson WS, Fletcher VR, Chenera B, Bartlett WR, Tham FS, Kullnig RK (1993) J Am Chem Soc 115:497. Guay D, Johnson WS, Schubert U (1989) J Org Chem 54:4731
8. Canonne P, Boulanger R, Bernatchez M (1987) Tetrahedron Lett 28:4997. Bailey WF, Ovaska TV (1990) Tetrahedron Lett 31:627. Chamberlin AR, Bloom SH, Cervini LA, Fotsch CH (1988) J Am Chem Soc 110:4788. Utimoto K, Imi K, Shiragami H, Fujikura S, Nozaki H (1984) Tetrahedron Lett 25:1999. Fujikura S, Inoue M, Utimoto K, Nozaki H (1985) Tetrahedron Lett 26:2101. Oppolzer W, Pitteloud R, Strauss HF (1982) J Am Chem Soc 104:6476. Bunce RA, Dowdy ED, Jones PB, Holt EM (1993) J Org Chem 58:7143 and references therein. Bunce RA, Harris CR (1992) J Org Chem 57:6981
9. Motherwell WB, Crich D (1991) Free radical reactions in organic synthesis. Academic Press, London . Curran DP (1991). In: Trost BM, Fleming I (eds) Comprehensive organic synthesis. Pergamon, Oxford, vol 4, chap 4.2. Mowbray CE, Pattenden G (1993) Tetrahedron Lett 34:127 and references therein. Hitchcock SA, Pattenden G (1992) Tetrahedron Lett 33:4843. Curran DP, Sisko J, Yeske PE, Liu H (1993) Pure Appl Chem 65:1153. Curran DP, Shen W (1993) Tetrahedron 49:755. Parker KA, Fokas D (1992) J Am Chem Soc 114:9688. Chen Y-J, Chen C-M, Lin W-Y (1993) Tetrahedron Lett 34:2961. Batey RA, Harling JD, Motherwell WB (1992) Tetrahedron 48:8031. Ozlu Y, Cladingboel DE, Parsons PJ (1993) Synlett 357. Grissom JW, Klingberg D (1993) J Org Chem 58:6559 and references therein
10. Davies HML, Clark TJ, Smith HD (1991) J Org Chem 56:3817 and references therein. Padwa A, Fryxell GE, Zhi L (1990) J Am Chem Soc 112, 3100 and references therein. Kim OK, Wulff WD, Jiang W, Ball RG (1993) J Org Chem 58:5571
11. Markó IE, Seres P, Swarbrick TM, Staton I, Adams H (1992) Tetrahedron Lett 33:5649 and references therein. Guevel R, Paquette LA (1994) J Am Chem Soc 116:1776 and references therein. Jisheng L, Gallardo T, White JB (1990) J Org Chem 55:5426. Sunitha K, Balasubramanian KK, Rajagopalan K (1985) Tetrahedron Lett 26:4393. Wender PA, Sieburth SM, Petraitis JJ, Singh SK (1981) Tetrahedron 37:3967 and references therein
12. Oppolzer W (1980) Heterocycles 14:1615. Funk RL, Vollhardt KPC (1980) Chem Soc Rev 9:41. Kametani T, Nemoto H (1981) Tetrahedron 37:3. Danheiser RL, Gee SK, Perez JJ (1986) J Am Chem Soc 108:806. Geiger JB, Eberbach W (1982) Tetrahedron Lett 23:4665. Nicolaou KC, Petasis NA (1984). In: Lindberg T (ed) Strategies and tactics in organic synthesis. Academic Press, New York, vol 1, pp 155–173. Denmark SE, Thorarensen A (1996) Chem Rev 96:137
13. Trost BM (1990) Acc Chem Res 23, 24. Trost BM, Shi Y (1993) J Am Chem Soc 115:12491
14. Collman JP, Hegedus LS, Norton JR, Finke RG (1987) Principles and applications of organotransition metal chemistry. University Science Books, Mill Valley

15. Overman LE, Abelman MM, Kucera DJ, Tran VD, Ricca DJ (1992) Pure Appl Chem 64:1813
16. Owczarczyk Z, Lamaty F, Vanter EJ, Negishi E (1992) J Am Chem Soc 114:10091
17. Grigg R, Sridharan V, Sukirthalingam S (1991) Tetrahedron Lett 32:3855
18. Meyer FE, Henniges H, De Meijere A (1992) Tetrahedron Lett 33:8039
19. van der Baan JL, van der Heide TAJ, van der Louw J, Klumpp GW (1995) Synlett 1
20. Snider BB, Vo NH, Foxman BM (1993) J Org Chem 58:7228 and references therein. Iwasawa N, Funahashi M, Hayakawa S, Narasaka K (1993) Chem Lett 545. Ali A, Harrowven DC, Pattenden G (1992) Tetrahedron Lett 33:2851. Grigg R, Kennewell P, Teasdale AJ (1992) Tetrahedron Lett 33:7789 and references therein. Grigg R, Kennewell P, Teasdale AJ (1992) Tetrahedron Lett 33:7789 and references therein. Batty D, Crich D (1992) J Chem Soc Perkin Trans 1 3205 and references therein. Trost BM, Shi Y (1992) J Am Chem Soc 114:791 and references therein
21. Adams J, Spero DM (1991) Tetrahedron 47:1765. Doyle MP (1986) Chem Rev 86:919. Ye T, McKervey A (1994) Chem Rev 94:1091
22. Padwa A (1991) Acct Chem Res 24:22. Padwa A, Krumpe KE (1992) Tetrahedron 48:5385
23. Osterhout MH, Nadler WR, Padwa A (1994) Synthesis 123
24. Hamaguchi M, Ibata T (1974) Tetrahedron Lett 4475
25. Gillon A, Ovadia D, Kapon M, Bien S (1982) Tetrahedron 38:1477
26. Padwa A, Hertzog DL, Chinn RL (1989) Tetrahedron Lett 30:4077.
27. Padwa A, Hertzog DL (1993) Tetrahedron 49:2589
28. Hertzog DL, Austin DJ, Nadler WR, Padwa A (1992) Tetrahedron Lett 33:4731
29. Maier ME, Evertz K (1988) Tetrahedron Lett 1677.
30. Maier ME, Schöffling B (1989) Chem Ber 122:1081
31. Doyle MP, Dorow, RL, Terpstra, JW, Rodenhouse, RA (1986) J Org Chem 51:4077
32. Sato M, Kanuma N, Kato T (1982) Chem Pharm Bull 30:1315
33. Regitz M, Hocker J, Leidhergener A (1973) In: Organic synthesis. John Wiley, New York, collect vol 5, pp 179–183
34. Hamaguchi M, Ibata T (1975) Chem Lett 499
35. Doyle MP, Pieters RJ, Tauton J, Pho HQ, Padwa A, Hertzog DL, Precedo L (1991) J Org Chem 56:820
36. Sundberg RJ, Pearce BC (1985) J Org Chem 50:245
37. Hill RK (1967). In: Manske RHF (ed) The alkaloids. Academic Press, New York, vol 9, p 483. Dyke SF, Quessy SN (1981). In: Rodrigo RGA, (ed) The alkaloids. Academic Press, New York, vol 18, p 1
38. Mondon A, Hansen KF, Boehme K, Faro HP, Nestler HJ, Vilhuber HG, Böttcher K (1970) Chem Ber 103:615. Mondon A, Seidel PR (1971) Chem Ber 104:2937. Mondon A, Nestler HJ (1979) Chem Ber 112:1329
39. Haruna M, Ito K (1976) J Chem Soc Chem Commun 345
40. Ishibashi H, Sato K, Ikeda M, Maeda H, Akai S, Tamura Y (1985) J Chem Soc Perkin Trans 1 605
41. Dean RT, Rapoport HA (1978) J Org Chem 43:4183
42. Stork G, Kretchmer RA, Schlessinger RH (1968) J Am Chem Soc 90:1647. Stork G (1968) Pure Appl Chem 17:383
43. Heathcock CH, Kleinman EF, Binkley ES (1982) J Am Chem Soc 104:1054. Heathcock CH, Kleinman EF (1979) Tetrahedron Lett 4125. Heathcock CH, Kleinman EF, Binkley ES (1978) J Am Chem Soc 100:8036
44. Barton DHR, McCombie SW (1975) J Chem Soc Perkin Trans 1 1574
45. Dehaen W, Hassner A (1991) J Org Chem 56:896
46. Prajapat D, Bhuyan P, Sandhu JS (1988) J Chem Soc Perkin Trans I 607
47. Prajapart D, Sandhu JS (1988) Synthesis 342
48. Tsuge O, Veno K, Kanemasa S (1984) Chem Lett 285
49. Padwa A, Hertzog DL, Nadler WR (1994) J Org Chem 59:7072
50. Ibata T, Hamaguchi M, Kiyohara H (1975) Chem Lett 21
51. Jacobi PA, Kaczmarek CSR, Udodung UE (1987) Tetrahedron 43:5475
52. Zani CL, de Oliveria AB, Snieckus V (1987) Tetrahedron Lett 28:6561

53. Carte B, Kernan MR, Barrabee EB, Faulkner DJ, Matsumoto GK, Clardy J (1986) J Org Chem 51:3528
54. Hirota H, Kitano K, Komatsubara T, Takahishi T (1987) Chem Lett 2079
55. Jacobi A, Craig TA, Walker DG, Arrick BA, Frechette RF (1984) J Am Chem Soc 106:5585
56. Hamaguchi M (1978) J Chem Soc Chem Commun 247
57. Padwa A, Hornbuckle SH (1991) Chem Rev 91:263
58. Huisgen R, Scheer W, Mader H (1969) Angew Chem Int Ed Engl 8:602. Heine HW, Peavy R (1965) Tetrahedron Lett 3123. Padwa A, Hamilton L (1965) Tetrahedron Lett 4363
59. Vedjs E, West FG (1986) Chem Rev 86:941
60. Huisgen R (1963) Angew Chem Int Ed Engl 2:565. Huisgen R, Grashey R, Steingruber E (1963) Tetrahedron Lett 1441
61. Grigg R, Kemp J, Sheldrick G, Trotter J (1978) J Chem Soc Chem Commun 109. Grigg R, Kemp J (1980) Tetrahedron Lett 2461
62. Padwa A (ed) (1984) 1,3-Dipolar cycloaddition chemistry. Wiley-Interscience, New York
63. Baret P, Buffet H, Pierre JL (1972) Bull Soc Chim Fr 6:2493
64. Bartnik R, Mloston G (1984) Tetrahedron 40:2569
65. Mara AM, Singh O, Thomas EJ, Williams DJ (1982) J Chem Soc Perkin Trans I 2169
66. Padwa A, Dean DC (1990) J Org Chem 55:405. Padwa A, Dean DC, Osterhout MH, Precedo L, Semones, MA (1994) J Org Chem 59:5347
67. Clineschmidt BV, Martin GE, Bunting PR (1982) Drug Dev Res 2:123
68. Loo PA, Braunwalder AF, Williams M, Sills MA Eur (1987) J Pharmacol 135:261
69. McDonald JW, Silverstein FS, Johnston MV Eur (1987) J Pharmacol 140:359
70. Daniels R, Salerni OL (1960) Proc Chem Soc 286
71. Band IBM, Loyd D, Singer MIC, Wasson FI (1966) J Chem Soc Chem Commun 544
72. Lloyd D, Singer MIC (1971) J Chem Soc (C) 2939
73. Durr H, Hev G, Ruge B, Scheppers G (1972) J Chem Soc Chem Commun 1257
74. Jones MB, Maloney VM, Platz MS (1992) J Am Chem Soc 114:2163
75. Zugravescu I, Petrovanu M (1976) *N*-Ylid Chemistry. McGraw Hill, New York
76. Platz MS, Maloney VM (1990) In: Platz MS (ed) Kinetics and spectroscopy of carbenes and biradicals. Plenum, New York, pp 239–252
77. Chen N, Jones M, White WR, Platz MS (1991) J Am Chem Soc 113:4981. Barcus RL, Hadel LM, Johnston LJ, Platz MS, Savino TG, Scaiano JC (1986) J Am Chem Soc 108:3928. Hadel LM, Platz MS, Scaiano JC (1983) Chem Phys Lett 97:446. Barcus RL, Wright BB, Platz MS, Scaiano JC (1983) Tetrahedron Lett 24:3955
78. Jackson JE, Soundararajan N, Platz MS, Doyle MP, Liu MTH (1989) Tetrahedron Lett 30:1335. Jackson JE, Soundararajan N, White W, Liu MTH, Bonneau R, Platz MS (1989) J Am Chem Soc 111:6874. White WR, Platz MS, Chen N, Jones M Jr (1990) J Am Chem Soc 112:7794. Perrin HM, White WR, Platz MS (1991) Tetrahedron Lett 32:4443. Morgan S, Jackson JE, Platz MS (1991) J Am Chem Soc 113:2782. Jones MB, Platz MS (1991) J Org Chem 56:1694. Sugiyama MH, Celebi S, Platz MS (1992) J Am Chem Soc 114:966.
79. Bonneau R, Liu MTH, Rayez MT (1989) J Am Chem Soc 111:5973. Liu MTH, Bonneau R (1989) J Phys Chem 93:7298
80. Chateaunef JE, Johnson RP, Kirchoff MM (1990) J Am Chem Soc 112:3217
81. Moss RA, Ho GJ, Shen S, Krogh-Jespersen K (1990) J Am Chem Soc 112:1638. Moss RA, Ho GJ (1990) Tetrahedron Lett 1225
82. Moss RA, Turro NJ (1990) In: Platz MS (ed) Kinetics and spectroscopy of carbenes and biradicals. Plenum, New York pp 213–238
83. Ho GJ, Krogh-Jespersen K, Moss RA, Shen S, Sheridan RS, Subramanian R (1989) J Am Chem Soc 111:6875
84. White WR III, Platz MS (1992) J Org Chem 57:2841
85. Rieser J, Friedrich K (1976) Liebigs Ann Chem 666
86. Adams J, Spero DM (1991) Tetrahedron 47:1765. Doyle MP (1986) Chem Rev 86:919. Padwa A, Krumpe KE (1992) Tetrahedron 48:5385. Ye T, McKervey A (1994) Chem Rev 94:1091
87. Zugrăvescu I, Rucinschi E, Surpăteanu G (1970) Tetrahedron Lett 941

88. Jackson JE, Soundararajan N, Platz MS (1988) J Am Chem Soc 110:5595
89. Padwa A, Austin DJ, Precedo L, Zhi L (1993) J Org Chem 58:1144
90. Kulyk MS, Neckers DC (1983) J Org Chem 48:1275
91. Connell R, Scavo F, Helquist P, Akermark B (1986) Tetrahedron Lett 27:5559
92. Doyle MP, Buhro WE, Davidson JG, Elliott RC, Hoekstra JW, Oppenhuizen M (1980) J Org Chem 45:3667
93. Turchi IJ, (ed) (1986) Oxazoles. Wiley Interscience, New York
94. Huisgen R, Sturm HJ, Binsch G (1964) Chem Ber 97:2865
95. Buu NT, Edward JT (1972) Can J Chem 50:3730. Shi G, Xu Y (1989) J Chem Soc Chem Commun 607
96. Doyle KJ, Moody CJ (1994) Tetrahedron 50:3761. (1994) Synthesis 1021
97. Connell R, Tebbe M, Helquist P, Akermark B (1991) Tetrahedron Lett 32:17
98. Yoo SK (1992) Tetrahedron Lett 33:2169
99. Doyle KJ, Moody CJ (1992) Tetrahedron Lett 33:7769
100. Gangloff AR, Akermark B, Helquist P (1992) J Org Chem 57:4797
101. Connell R, Tebbe M, Gangloff AR, Helquist P, Akermark B (1993) Tetrahedron 49:6446
102. Kende AS, Hebeisen P, Sanfilippo PJ, Toder BH (1982) J Am Chem Soc 104:4244
103. Sheppard WA, Gokel GW, Webster OW, Betterton K, Timberlake JW (1979) J Org Chem 44:1717
104. Grigg R, Ardill H, Sridharan V, Surendrakumar S, Thianpatanagul S, Kanajun S (1986) J Chem Soc Chem Commun 602. Grigg R (1987) Chem Soc Rev 16:89
105. Burger K, Schickaneder H, Zettle C (1977) Angew Chem Int Ed Engl 16:54. Burger K, Schinkander H, Zeittl C (1982) Justus Liebigs Ann Chem 1730
106. Wentrup C (1978) Helv Chim Acta 61:1755
107. Sha CK, Young J (1984) Heterocycles 22:2571. Liu JM, Young JJ, Li YJ, Sha CK (1986) J Org Chem 51:1120. Sha CK, Ougang SL, Hsieh DY, Chang RC, Chang SC (1986) J Org Chem 51:1490
108. Padwa A, Caruso T, Nahm S, Rodriguez R (1982) J Am Chem Soc 104:2864. Padwa A, Dent WH, Schoffstall AM, Yeske PE (1989) J Org Chem 54:4430
109. Nenitzecu CD (1968) In: Olah G, Schleyer P von R, (eds) Carboniumions. Wiley Interscience, New York, vol I, chap 1, p 1
110. Saunders M, Chandrasekhar L, Schleyer P von R (1980) In: de Mayo P (ed) Rearrangement in ground and excited states. Academic Press, New York, vol I
111. Brouwer DM, Hogeveen H (1972) Prog Phys Org Chem 9:179
112. Kirmse W (1971) Carbene chemistry, 2nd edn. Academic Press, New York
113. Jones WM (1980) In: de Mayo P (ed) Rearrangement in ground and excited states. Academic Press, New York, vol I
114. Walling C (1963) In: de Mayo P (ed) Molecular rearrangements. Wiley Interscience, New York, pt I
115. Julia M (1971) Acc Chem Res 4:386
116. Beckwith AL, Ingold KU (1980) In: de Mayo P (ed) Rearrangement in ground and excited states. Academic Press, New York, vol I
117. Padwa A, Dean DC, Zhi L (1992) J Am Chem Soc 114:593. Padwa A, Dean DC, Lin Z (1989) J Am Chem Soc 111:6451
118. Rodgers JD, Caldwell GW, Gauthier AD (1992) Tetrahedron Lett 33:3273
119. Padwa A, Zhi L (1990) J Am Chem Soc 112:2037. Padwa A, Price AT, Zhi L (1996) J Org Chem 61:2283
120. Padwa A, Price AT (1995) J Org Chem 60:6258
121. Cordell GA (1979) In: Manske RHF, Rodrigo RGA (eds) The alkaloids. Academic Press, New York, vol 17, pp 199–384. Saxton JE (1993) Nat Prod Rep 10:349–395
122. Regitz M (1967) Angew Chem Int Ed Engl 6:733. Taber DF, Ruckle RE Jr, Hennessy MJ (1986) J Org Chem 51:1663
123. Moza BK, Troja'nek, J (1963) Collect Czech Chem Commun 28:1427. Büchi G, Matsumoto KE, Nishimura H (1971) J Am Chem Soc 93:3299

Author Index Volumes 151–189

Author Index Vols. 26–50 see Vol. 50
Author Index Vols. 51–100 see Vol. 100
Author Index Vols. 101–150 see Vol. 150

The volume numbers are printed in italics

Springer and the environment

At Springer we firmly believe that an international science publisher has a special obligation to the environment, and our corporate policies consistently reflect this conviction.

We also expect our business partners – paper mills, printers, packaging manufacturers, etc. – to commit themselves to using materials and production processes that do not harm the environment. The paper in this book is made from low- or no-chlorine pulp and is acid free, in conformance with international standards for paper permanency.

Printing: Saladruck, Berlin
Binding: Buchbinderei Lüderitz & Bauer, Berlin